THE VIRUS
A History of the Concept

THE VIRUS
A History of the Concept

Sally Smith Hughes

Heinemann Educational Books
London
Science History Publications
New York • 1977

First published in the United States by
Science History Publications
a division of
Neale Watson Academic Publications, Inc.
156 Fifth Avenue, New York 10010

(CIP data on final page)

Published in Great Britain by
Heinemann Educational Books Ltd.
48 Charles Street, London W1X 8AH

LONDON EDINBURGH MELBOURNE AUCKLAND TORONTO
HONG KONG SINGAPORE KUALA LUMPUR NEW DELHI
NAIROBI JOHANNESBURG LUSAKA IBADAN

ISBN 0 435 54755 0

First published 1977

cover photograph: A rare scanning electron micrograph showing a viral attack of T2-coliphages adsorbed on *E. coli* B bacteria. Magnification: 1000 A = 12 mm. (Photo by Dr. Barbara Panessa, N.Y.U. Medical Center, New York and Dr. Alec Broers, Thomas J. Watson Research Center, IBM, Yorktown Heights, N.Y.)

cover design: Ellen Tedesco Boardman

Printed in the U.S.A.

Contents

To my father

Disease is from old and nothing about it has changed. It is we who change, as we learn to recognize what was formerly imperceptible.

Jean-Martin Charcot

A virus, consisting of genetic material enclosed in a protective coating, is one of the simplest entities able to reproduce. Viruses have no metabolic systems, they have no intrinsic motility, they cannot respond to stimuli, and they do not grow in the usual sense. The ability to maintain genetic continuity, with the possibility for mutation, is the only basis for considering viruses to be alive.

Furthermore, the mechanism by which viruses reproduce is unique in biology. This seems to be true whether one is talking about a virus of green plants, bacteria, mushrooms, algae, insects, higher animals, or humans. In all cases, during the reproductive cycle the genetic material of the virus becomes a functional part of the cell it has infected. The genes added to the cell by the virus cause the infected cell either to produce more virus particles, with cell death usually the end result, or to become changed and acquire new characteristics. (*Goodheart*, 1969, p. 1)

Notes on Terminology and Translations

The word 'micro-organism' as used in this book applies to all living organisms of light microscopic size; it does not apply to viruses. The term 'infectious agent' applies to any microscopic or submicroscopic micro-organism or virus. Virology was not recognized as an independent science before 1950. Therefore, 'viral research' has been used in preference to 'virology' or 'virological research' to refer to work on viral diseases prior to that date. The term 'virus' since antiquity has been synonymous with 'poison,' and more recently with 'infectious agent.' (See Appendix I for a history of the word 'virus.') Some quotations contain the word 'virus' used in the latter sense of 'infectious agent' rather than to designate the unique entity which today is specified by this word. When 'virus' is used with the more general meaning of 'infectious agent,' these two words are inserted in brackets in the quotation. The adjective 'filterable,' as in 'filterable virus,' is used to signify that the substance in question (in the above example, the virus) passes through the filter.

Translations, unless otherwise stated, are those of the present author. Italics occurring within quotations are always those of the original author.

Sources and References

This book is based almost exclusively on primary source material. For reading ease, references are not usually given in the text. Instead they will be found under the author's name in alphabetical order in the bibliography which also includes general references. Page numbers for quotations or specific points appear at the end of the relevant reference. It is only when there is danger of confusion that the page number and/or date are noted in parentheses in the text.

Acknowledgements

I am grateful to Professor A.P. Waterson of the Royal Postgraduate Medical School, London and to Dr. Edwin Clarke of the Wellcome Institute for the History of Medicine, London for their guidance in the preparation of the doctoral dissertation upon which this book is partially based. I am most indebted to Trevor and also to Drs. Steven Bachenheimer, Andrew Cunningham, Seymour Mauskopf, Michael McVaugh, and to Martha Dill for discussion and criticism of various aspects of this book. The research for the doctoral dissertation was made possible by a research fellowship from the Wellcome Trust, London.

Preface

This history of the concept of the virus is written in nontechnical language for those with a general interest in science as well as for virologists and historians of science and medicine. Considering the clinical, theoretical and social importance of the viruses, it is surprising that so little of their history has appeared in print.*

Little emphasis has been placed in this book on dating the 'discovery' of the virus for two reasons. First, there is the problem of determining when it was discovered. Of the following, which should one designate? Its recognition as a filterable, submicroscopic infectious agent; its visualization in the light microscope as inclusion or elementary body; its subsequent visualization in the electron microscope; the elucidation of its molecular structure and function; or the enunciation of the earliest hypothesis approximating most closely its fundamental properties?

Second, the development of the present concept has been a cumulative one. Therefore, it would be misleading to point to one individual as the 'discoverer' of the virus, as has occurred in the case of Ivanovski and Beijerinck.

Others have credited these men with the 'discovery' of the virus. As we shall see in the chapter devoted to their work, Ivanovski's contribution was to demonstrate for the first time the existence of filterable infectious agents. Beijerinck's was to propose a concept of the virus which in some respects is remarkably similar to the one we have today. The influence of their contributions will be considered.

Until one recognizes that infectious diseases are caused by micro-organisms, there is little reason to suspect the existence of even smaller infectious agents, the viruses. It follows that to understand the development of the concept of the virus, one must establish the microbiological context from which it sprang. Under the influence of the germ theory, bacteriological methods and

*Aside from the short historical introductions appearing in assorted textbooks and papers on virological subjects, Robert Doerr's chapter entitled "Die Entwicklung der Virusforschung und ihre Problematik" (in *Handbuch der Virusforschung*, R. Doerr and C. Hallauer, Vienna: Julius Springer, 1938, vol. 1, 1st half, pp. 1–21) and a chapter in *Three Centuries of Microbiology* (H.A. Lechevalier and M. Solotorovsky, New York: McGraw-Hill, 1955, pp. 280–332) consisting of excerpts from papers relevant to the history of virology with comments by the authors, are the only historical works on virology, to the present author's knowledge, that have yet been published in book form.

techniques were applied in the initial recognition of filterable, submicroscopic infectious entities and in the early years of twentieth century viral research. Chapter one describes the development of the germ theory of infectious disease.

The second chapter deals with the next step towards the eventual formulation of the modern concept of the virus: the application of bacteriological methods to the investigation of animal and plant diseases. As these methods became increasingly sophisticated, diseases were recognized which were obviously infectious but in which no causal micro-organisms could be identified.

The third chapter is concerned with the realization that not all known agents of infectious disease share the bacterial properties of visibility in the light microscope, retention by bacterial filters, and cultivability in artificial media. Moreover, a few scientists proposed theories about the nature of infectious agents which conflicted with the germ theory. These theories raised the possibility of the existence of infectious agents which were not micro-organisms.

The fourth chapter describes Ivanovski's research on tobacco mosaic disease whose cause, he reported, is a filterable infectious agent, and Beijerinck's rediscovery of the same agent and hypothesis of the contagium vivum fluidum.

The fifth chapter traces the development of the microbial and the nonmicrobial concepts of the virus up to 1900 and includes the work of Loeffler and Frosch, Nocard and Roux, Sanarelli, Koning, Woods, and Heintzel.

The description of the evolution of these concepts in the twentieth century is covered in chapter six. Most investigators in the early decades favored the microbial theory and hence regarded the virus as a submicroscopic, filterable microbe. A minority believed it to be an inanimate substance. Neither theory adequately described the intrinsic properties of the viruses. After 1950 experimental techniques, which had been developed and perfected for virological research, allowed their unique biochemical composition, physical structure, and mechanism of replication within the host cell to be elucidated.

In this book every attempt has been made to avoid straying too far into philosophy, metaphysics or sociology when discussing the implications of the concept of the virus. However, certain fundamental questions involved in virological research have been mentioned, such as the definition of life and nonlife, and the origin of viruses.

An extensive bibliography is included in the hope that this book might serve as a starting point for the exploration of other facets of this fascinating subject—the history of virology.

1

The Germ Theory of Infectious Disease

Supernatural and cosmic phenomena were originally deemed to be the cause of infectious diseases. As man gained knowledge of the natural world, these beliefs were replaced by the idea that infectious substances produced diseases in living organisms. By the early part of the nineteenth century there were three main theories concerning the nature of the infectious agent: the contagionist, the zymotic and the miasmatic.

The contagionist theory held that infectious diseases were caused by particles which reproduced in the body. Some contagionists believed that the particles, although capable of reproduction, were inanimate; others, the animaculists, thought that they were minute living organisms.[1]

The second theory of infectious disease, the zymotic, was actually a nineteenth-century variant of the contagionist theory. It was based on an analogy between the processes of infection and fermentation. Fermentation was a subject of particular interest at this time. The zymotic theory assigned to the infectious material the properties of a ferment or 'zyme' specific to each disease. This zyme multiplied only within living organisms where it produced the zymotic disease in question.

The third theory, the miasmatic, was promoted by the anticontagionists who thought that noxious gases or miasmata arising from the ground or from decaying animal and vegetable matter could produce disease in individuals exposed to them.

In the second half of the nineteenth century evidence for the pathogenicity of micro-organisms accumulated. This evidence resulted in the germ theory of infectious disease, a more scientific version of the contagionist theory, which attributed the cause of infectious diseases to micro-organisms. With increasing confidence in the germ theory, the zymotic and miasmatic theory lost favor.

The following is not a detailed history of the germ theory[2] but an account of its development intended to set the stage for discussion of the virus. This is necessary because the germ theory provided the theoretical context for the conceptualization of yet smaller particles, the viruses.

The Formulation of the Germ Theory in the Early Nineteenth Century

The theoretical and experimental work which culminated in the confirmation and acceptance of the germ theory in the 1870s and 1880s stemmed from two related areas of scientific endeavor in the first half of the century: investigations of infectious diseases and investigations of fermentation.

Investigations of Infectious Diseases. Agostino Bassi, an early protagonist of the germ theory, was the first to demonstrate the causal relationship between a specific micro-organism and a specific disease. In 1835 he attributed the silkworm disease of muscardine to "an external organism, which enters the insect and produces the disease, death, and the subsequent hardening and efflorescence of the corpse." "This murderous creature," he continued, "is organic, living, and vegetable. It is a cryptogamic plant, a parasitic fungus."[3] He subsequently extended his theory of infection by micro-organisms to certain diseases of man.

The microscopic examination of infected animal tissues for pathogenic micro-organisms was actively pursued in the decade following Bassi's discovery, but few positive contributions were made, partly because of the poor resolving powers of current microscopes and the absence of sophisticated fixing and staining techniques. The few observations made at this time concerned micro-organisms which were readily visible in unstained preparations, such as Johann Lucas Schönlein's discovery of the fungus which causes the skin disease of favus and John Goodsir's observation of *Sarcina ventriculi*, a type of micrococcus which forms large cuboidal groups. More rapid advances were made in plant pathology where fungi are the major cause of infectious disease. By the 1840s microscopic fungi were under active investigation as plant pathogens and were found to be responsible for many rusts, smuts and other plant diseases. Thus long before the germ theory had been generally accepted, plant pathologists had acknowledged that certain plant diseases are caused by micro-organisms.

A publication of the early nineteenth century which has particular relevance to the history of the germ theory of disease is Jacob Henle's "Von den Miasmen und Contagien" of 1840. Rather than being a presentation of original experimental findings, the paper is an assessment of seemingly unrelated information about micro-organisms, including Bassi's work on muscardine and the recent research on fermentation, which had been gathered by investigators in diverse areas of research. Knowing that infectious diseases could be transmitted along a chain of individuals, Henle concluded that the causal agents must multiply within the bodies of infected organisms. This capacity to reproduce as well as to assimilate foreign materials—properties which he associated exclusively with living beings—convinced Henle that

> . . . the matter of the contagions is not only organic, but also animate, indeed endowed with individual life, and . . . stands in the relationship of a parasitic organism to the diseased body.

In addition to promoting the doctrine of animate contagion, Henle indicated principles which might guide future experimental research on infectious disease. He recognized the need, in establishing the etiology of an infectious disease, to demonstrate the constant association of a specific micro-organism with a specific disease and to demonstrate also that the microbe, after isolation from the material containing it, was still able to cause the disease.[4]

"Von den Miasmen und Contagien" was greeted hesitantly by scientists of the 1840s, probably because it had a theoretical rather than an experimental basis. In the wake of the speculative theories of *Naturphilosophie* they were distrustful of hypothesis, however cogent. Henle was criticized for drawing analogies between diseases of lower animals and those of man, especially in the absence at that time of proof for the existence of micro-organisms which cause infections in higher animals. Furthermore, current experimental techniques, as Henle himself admitted, were incapable of satisfying his criteria for proving an etiology.

Investigations of Fermentation. The second broad area of research which provided evidence for the implication of micro-organisms in natural processes came from studies of fermentation. In the years 1837 and 1838 Charles Cagniard-Latour, Theodor Schwann and Friedrich Traugott Kützing independently published experimental results indicating that yeast, contrary to the opinion of many of their contemporaries, was a living organism and that it was responsible for the fermentation of sugar. Hence fermentation in their view was the result of microbial activity rather than the strictly chemical reaction proposed by Jons Jacob Berzelius, Friedrich Wöhler, Justus von Liebig and others.

Despite evidence from many sources of the participation of micro-organisms in fermentation and in certain infectious diseases, the lack of conclusive evidence due to obstacles of both a theoretical and a technical nature delayed widespread acceptance of the germ theory of disease until after 1877. The most serious obstacle was the lack of conclusive evidence for the relevance of the germ theory to infectious diseases of higher animals and man. Bassi's work on muscardine, observations of micro-organisms found in association with human diseases, and the phytopathological discoveries of the early mycologists could all be dismissed as inapplicable to the great majority of human and animal infections. The settlement of the controversy between miasmatists and contagionists depended upon the development of exact techniques for isolating and identifying the causal agents of specific diseases. This development did not take place until the late 1870s.

Development of the Germ Theory after 1850

After 1850 a gradual reorientation of thought about infectious diseases occurred. New evidence was reported for the implication of micro-organisms in biological processes. Louis Pasteur's work on fermentation in the 1850s and 1860s represents a crucial step in establishing the germ theory.[5] He developed the ideas, proposed earlier in the century but not widely accepted, of fermentation and putrefaction as microbial processes. He showed that the fermentation of wine, vinegar and beer, and the putrefaction of organic infusions were the result of the activity of specific micro-organisms. Pasteur also demonstrated that the "diseases" of wine, vinegar and beer were due to the presence of alien micro-organisms which changed the chemical and physical properties of these fluids.[6] The germ theory of fermentation eventually displaced the competing idea of fermentation by chemical catalysis.

The controversy over the spontaneous generation of micro-organisms, a question dating back to classical times, was another area of scientific concern. The fact that micro-organisms are generated only by micro-organisms, not *de novo*, had to be established before the germ theory and the related doctrine of the specificity of infection could be accepted. If micro-organisms arose spontaneously, then they possibly could be by-products rather than the cause of the diseases with which they were associated. It was the proponents of spontaneous generation who were the main opposition to the germ theory.

In a series of elegant experiments performed in the late 1850s and 1860s, Pasteur demonstrated that micro-organisms did not appear in sterile media and therefore did not arise spontaneously under normal laboratory conditions. In 1876 the British physicist John Tyndall strengthened the argument against spontaneous generation by demonstrating that sterile infusions remained sterile in air shown by a beam of light to be "optically inactive," while media exposed to "optically active" air containing dust and "germs" showed bacterial growth. His observation that prolonged heating was necessary to sterilize some infusions led him to believe that bacteria had different developmental forms. This suspicion was confirmed later in the same year when the German botanist Ferdinand Cohn discovered the heat-resistant bacterial spore.

In the mid-1860s the works of Pasteur on the chemical activities of bacteria and on spontaneous generation came to the attention of the British surgeon Joseph Lister. Recognizing the similarity between fermentation and putrefaction in wounds, Lister applied Pasteur's ideas to problems encountered in surgical procedures. Concluding that bacteria, which Pasteur had shown to be ubiquitous in the air, were the cause of suppuration of wounds, Lister tested a number of chemical substances for their ability to sterilize air. He finally chose carbolic acid spray which he used during operations to destroy airborne germs; he succeeded in significantly reducing the number of fatalities due to surgical infections. The age of antiseptic surgery, based on the germ theory of disease, had begun.

Pasteur turned for the first time to research on an animal disease when in 1865 he was asked to investigate an epidemic disease of silkworms which threatened to ruin the silk industry in France. By 1870 he had identified two diseases, pébrine and flacherie, both of which he attributed to micro-organisms.[7]

Pasteur was not the only investigator to provide evidence in the 1860s for a germ theory of infection. For example, the French veterinarian Jean Baptiste Auguste Chauveau demonstrated in a series of experiments that infectivity was a property of the particulate rather than the fluid components of vaccine lymph. His compatriot, Jean Antoine Villemin, found that tuberculosis could be transmitted by inoculation from humans to animals and successively from animal to animal. He suggested that it was a specific disease caused by an infectious principle which multiplied like a parasite in the bodies of infected individuals. The English microscopist Lionel Beale attributed infectious properties to minute particles of living "germinal" matter which he believed were derived from the tissues of sick humans and animals. Despite the speculative basis of Beale's theory of infectious disease, his research on cattle plague in the 1860s gave currency to the concept of vital, particulate agents of infection. The work of Casimir-Joseph Davaine in the 1860s, unlike the hypotheses of Beale, was based on extensive experimentation. Davaine showed that anthrax was transmitted only when the inoculum contained the "bacteridia," today known as *Bacilli anthracis*.[8] However, he was unable to prove beyond doubt that the infectious agent was not a soluble toxin accompanying the bacteria or to explain its survival for long periods of time in a desiccated state. The search for pathogenic micro-organisms continued in the 1870s with workers such as Edwin Klebs, Otto Obermeier and Carl Weigert identifying specific bacteria in diseased tissues, although they were unable to prove a causal relationship.

The publication in 1877 of Koch's research on anthrax provided crucial evidence in support of the germ theory of disease. He, like Davaine, found that the bacillus was always present in the tissues of animals suffering from anthrax. Through continuous microscopical examination of blood drawn from infected mice and close observation of cultures of the bacilli, he elucidated the complete life cycle of the bacillus, both inside and outside the host. This was important evidence against the current notion that bacteria represented early stages in the development of higher forms of life. Koch's discovery that the bacillus possessed spores which remained virulent for years explained the survival of the pathogen under adverse conditions, thereby answering a question unanswered by Davaine. By inoculating animals with cultures of the bacillus, Koch produced a disease with the symptoms of anthrax and demonstrated that its experimental production was invariably associated with the presence of the anthrax bacillus or its spores. The techniques used by Koch in this study were soon to provide the experimental basis for ascertaining the specific bacterium causing a disease.

Later in the same year (1877) Pasteur and Joubert established beyond reasonable doubt that anthrax was caused by the bacillus, rather than by a soluble toxin accompanying it. They showed that filtrates of anthrax-infected blood failed to produce the disease in experimental animals. With this and other experiments Pasteur lent further weight to the germ theory of disease whose confirmation may be dated from these investigations of 1877.

Table I

LANDMARKS OF BACTERIOLOGY (1877–1900)*

Date	Scientist	Accomplishment
1877	R. Koch L. Pasteur	Independent elucidation of the etiology of anthrax; the first widely-accepted demonstration of the bacterial cause of a disease.
1877	T.J. Burrill	Discovery of the bacterial etiology of a plant disease (pear blight).
1877	L. Pasteur	Discovery of the bacillus of malignant edema.
1879	A. Neisser	Discovery of the gonococcus.
1880	L. Pasteur G.M. Sternberg	Independent discovery of the pneumococcus.
1880	L. Pasteur	Discovery of the bacillus of fowl cholera.
1880	L. Pasteur	Discovery of staphylococci in pus.
1880	A. Hansen	Discovery of the bacillus of leprosy.
1880	C.J. Eberth R. Koch	Discovery of the bacillus of typhoid.
1881	A. Ogston R. Koch	Independent discovery of the streptococcus.
1882	R. Koch	Isolation of the bacillus of tuberculosis.
1882	F. Loeffler and W. Schütz	Isolation of the bacillus of glanders.

*Primarily based on Bulloch's *History of Bacteriology*.

1884	R. Koch	Isolation of the vibrio of cholera.
1884	F. Loeffler	Isolation of the bacillus of diphtheria.
1884	J. Rosenbach	Isolation of the staphylococcus and streptococcus.
1884	G. Gaffky	Isolation of the bacillus of typhoid.
1884	R. Koch	Isolation of the cholera vibrio.
1884	A. Nicolaier	Discovery of the bacillus of tetanus.
1884	F. Loeffler	Isolation of the bacillus of diphtheria.
1885	E. von Bumm	Isolation of the gonococcus.
1886	A. Fräenkel	Isolation of the pneumococcus.
1887	A. Weichselbaum	Isolation of the meningococcus.
1888	A. Gärtner	Discovery of *Bacillus enteritidis* and *paratyphosus*.
1889	S. Kitasato	Isolation of the bacillus of tetanus.
1892	W.H. Welch and G.H.F. Nuttall	Isolation of the bacillus of gas gangrene.
1894	S. Kitasato and A. Yersin	Isolation of the bacillus of plague.
1897	E.P.M. van Ermengem	Isolation of the bacillus of botulism.
1898	K. Shiga	Isolation of the bacillus of dysentery.

Its acceptance was followed by an era of great productivity in medical bacteriology, during which the germ theory was applied to research on various infectious diseases. The bacteriological methodology and experimental techniques which were developed and refined facilitated the bacterial discoveries of the "golden age" of bacteriology. (See Table I) Along with these successes, modifications in scientific attitudes towards research on infectious disease and towards the nature of the infectious agent occurred. These changes will be considered in the next two chapters.

Notes

1. For the early history of the animalculist theory see Belloni, *Le "contagium vivum" avant Pasteur.*
2. Secondary sources on the germ theory:
   Bulloch, *The History of Bacteriology*;
   Ford, *Bacteriology*;

Foster, *A History of Medical Bacteriology and Immunology*; Rosen, *A History of Public Health*.

3. Bassi's causal agent was a microscopic fungus, later named in his honor, *Botrytis bassiana*.
4. Thirty years later Robert Koch, a pupil of Henle, applied these principles experimentally in a form which became popularized as 'Koch's postulates.' They are discussed in Chapter 2.
5. Pasteur's work on fermentation and the eventual acceptance of his theory of fermentation instead of the chemical interpretation of Berzelius, Liebig, Wöhler and others is discussed in Dubos, *Louis Pasteur: Free Lance of Science*.
6. Pasteur's papers on these topics have been collected in the following volumes of the *Oeuvres de Pasteur*: vol. 3, *Études sur le Vinaigre et sur le Vin*; vol. 4, *Études sur la Bière*.
7. L. Pasteur, *Oeuvres de Pasteur*, vol. 4. It is now known that pébrine is caused by a protozoon, *Nosema bombycis*, and that flacherie is caused by a virus which predisposes the silkworm to infection by the bacterium, *Bacillus bombycis*, which Pasteur had observed. (Lechevalier and Solotorovsky)
8. Davaine published many papers on anthrax between 1863 and 1870, most of which appeared in the *C. r. hebd. Séanc. Acad. Sci., Paris* or the *Bull. Acad. Méd.* An exhaustive discussion of Davaine's research on anthrax and the problems it raised is contained in "Un grand médecin et biologiste Casimir-Joseph Davaine (1812–1882)," by J. Théodoridès.

2

Microbiology in the Late Nineteenth Century

By the 1890s it was common knowledge in microbiology that the causal agents of a number of important infectious diseases remained unknown, despite extensive research. Infectious agents other than bacteria, such as the protozoa, were investigated and the existence of as yet unknown types of infectious agents became a distinct possibility.

For several reasons it is necessary to discuss the late nineteenth century microbiological background to these events. First, microbiological techniques had to become sufficiently sophisticated to allow investigators to exclude with confidence bacterial etiologies[1] for diseases which are now known to be viral. Second, the techniques for the discovery of filterable infectious agents in the last decade of the nineteenth century were derived from microbiology. Third, the impact of the discovery of these agents and of the subsequent growth of the viral concept can only be appreciated if the existing intellectual climate in microbiology is known. Fourth, the methods initially used to investigate filterable infectious agents were those which had been successfully applied in microbiology.

In this chapter the recognition of infectious diseases of unknown etiology is considered in the context of the intellectual climate of microbiology in the 1880s and '90s. The topics discussed illustrate the increasing scope of microbiological inquiry—an atmosphere conducive to the discovery of a new type of infectious agent.

Medical Bacteriology

In the bacteriological fervor of the years following acceptance of the germ theory, bacteria were assumed to be the cause of almost all human and animal infections. Sometimes bacteria which happened to be present in infectious materials were wrongly interpreted to be the cause of the disease in question. Even diseases later found to be nonbacterial, such as yellow fever and rickets, were initially given bacterial etiologies.

In the late 1880s the quality of bacteriological research rapidly improved when new techniques were applied using the guidelines for the isolation of the tubercle bacillus presented by Koch in 1882 and 1884. The three conditions which Koch believed necessary to establish a bacterial etiology later became known as "Koch's postulates." These are:

1. The specific microbe must be demonstrated in all cases of the disease.
2. It must be isolated and cultured in a pure state on an artificial medium.
3. The pure culture must produce the disease when inoculated into healthy, susceptible animals.

Although bacteriologists accepted these postulates on a theoretical basis, in practice they were not always easy to fulfill.

Some of the difficulties involved in applying Koch's postulates in the 1880s are illustrated in *Vaccinia and Variola*, the work of the Scottish physician John Brown Buist. At this time he was teaching the technique of vaccination under the auspices of the Local Government Board of Scotland. He was also doing research on vaccines to determine "the bacteric form in which the contagium of vaccinia and variola exists, in the materials which are capable of reproducing these diseases by inoculation." (p. vi) He gave detailed accounts and illustrations of the variously colored microbial colonies which he was able to grow by culturing different vaccines on gelatine or agar substrates. His description of the vaccines in terms of color and degree of opacity suggests that they and consequently his cultures were badly contaminated.

Buist drew directly from vaccinial or variolar pustules what he described as "clear lymph." This material was relatively uncontaminated by extraneous micro-organisms and, in contrast to the cultures which he grew from vaccines, was difficult to stain. However, he devised a procedure for fixing and staining the lymph which, under a high power microscope, revealed tiny spherical particles with a diameter of about 0.15 of a micron. (See Figs. 1 and 2) These he regarded as "spores, which develop by artificial cultivation in solid media into the larger [bacterial] forms. Clear vaccine and variolous lymph, therefore, contain spores of bacteria in suspension, which are distinguished from the

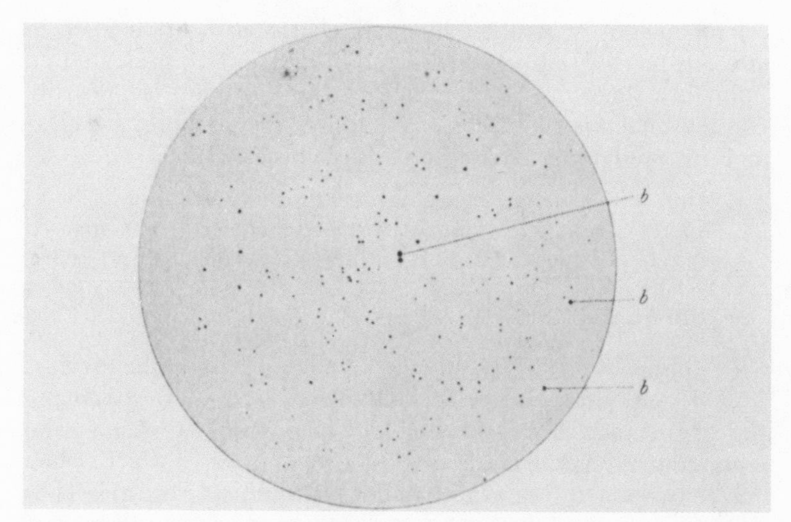

Fig. 1 Cover-glass preparation of clear vaccine lymph stained with aniline methyl violet. The elementary bodies, varying in size from 0.1 to 0.5 mu, are clearly visible. Magnification 1045 diameters.

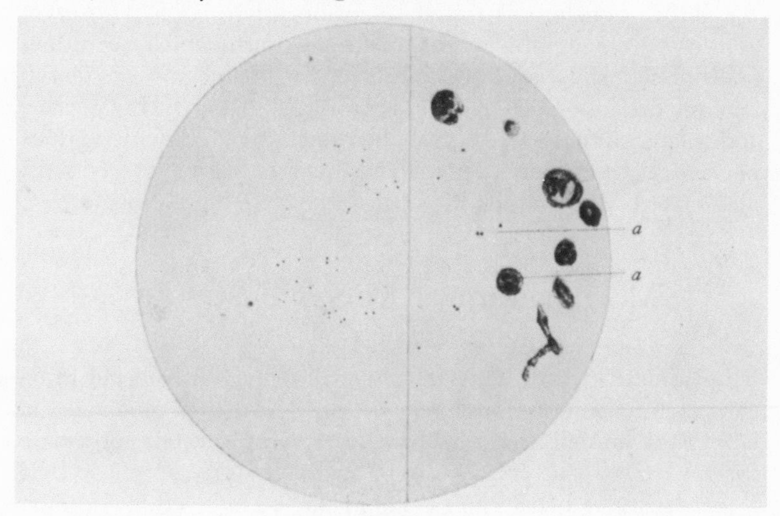

Fig. 2 Left half: Cover-glass preparation of clear variolous lymph stained with gentian violet. The elementary bodies vary in size from 0.2 to 0.3 mu. Right half: Opaque variolous lymph stained with gentian violet and chrysoidin, showing contamination by micrococci, diplococci and pus cells. Magnification 1045 diameters.

(Reproduction of Plates VIII and XI from Buist's *Vaccinia and Variola*, London: J. & A. Churchill, 1887.)

micrococci of cultivation by their characteristic difference in size." (pp. 76–77) In Buist's opinion, the spores represented an immature stage in the life cycle of the bacterium supposedly causing smallpox and vaccinia. He found, furthermore, that clear and opaque lymph differed in physiological action:

> While the materials containing very minute organisms [Buist's "spores"] produce local pocks, those containing or composed entirely of larger organisms [bacteria] produce either [an] imperfect local effect, or no local effect, but an eruption in some other part. (p. 159)

The "spores" were apparently the active principle of the vaccine.

Modern virologists believe that Buist's "spores" were actually the elementary bodies of smallpox and vaccinia, that is, the comparatively large viral particles of these diseases which, when correctly stained, are visible under the light microscope. Thus Buist in the year 1886 became the first to see and to develop a stain for viral particles. He did not of course recognize their true identity.

Buist's conclusions rested solely on microscopical observations of vaccine lymph and on artificial cultures which, according to Koch's postulates, were insufficient evidence for a bacterial etiology because they could not reproduce smallpox. His vaccines and plate cultures were probably contaminated with various micro-organisms and hence those he saw were not, as Buist maintained, stages in the life cycle of a single type of bacterium.

Modification of Koch's Postulates

It became evident in the 1880s that all diseases suspected to be caused by bacteria could not be produced experimentally in laboratory animals, nor could the artificial substrate appropriate for the growth of every bacterium be found. For example, in 1883 Koch discovered that cholera could not be induced in common experimental and domesticated animals, thereby precluding the fulfillment of the postulate for the artificial production of the disease in susceptible animals. He also knew that other infectious diseases of humans, such as leprosy and typhoid, had not been transmitted experimentally to animals. Faced with these obstacles, he stated in 1884 in a paper on cholera that a bacterium could be accepted as the cause of an infection, even though the disease had not been artificially produced in an experimental animal,

> if we can establish the constant presence of a certain kind of bacteria in the disease under consideration, and the absence of the same bacteria in other diseases.

This was equivalent to saying that his third postulate could not be satisfied in every instance. Several years later Koch stated this explicitly:

> If only the first two requirements of the proof are satisfied, if the constant and exclusive occurrence of the parasite is established, the causal connection between parasite and disease is validly proved. Starting from this basis, we must admit that a series of diseases in which the experimental infection of animals has so far failed or only been partially successful, thus leaving the third part of the proof unfulfilled, is nonetheless to be regarded as parasitic. (1890, p. 757)

Plant Bacteriology

Accounts of the early history of bacteriology seldom mention the fact that plant bacteriology did not emerge as a distinct speciality until almost the end of the nineteenth century. In fact, after acceptance of the germ theory, almost a quarter of a century passed before bacteria were commonly acknowledged as a cause of plant infections.

One reason for the delay was the dominant position of fungi in plant pathology, a position which they had occupied for more than thirty years at the time of the confirmation of the germ theory. (1877) Because of the importance of fungi as plant pathogens, there was at this time little incentive to search for other types of disease agents. Hence a body of evidence linking bacteria and plant diseases, similar to that which had accumulated by the 1870s for the association of bacteria with human and animal diseases, did not exist for plants. The misconception arose that only humans and animals were subject to bacterial diseases. This error persisted partly because research on pathogenic bacteria was dominated by medically oriented investigators who were not particularly interested in plant infections.

A second reason for the slow acceptance of the bacterial theory of disease in plant pathology was that botanists and mycologists doubted that bacteria were a significant cause of infections in plants.[2] For example, Anton de Bary, the "founder of mycology," remarked in 1885:

> According to present experience, parasitic bacteria are after all only of slight importance as contagia of plant diseases. Most contagia of the numerous infectious diseases of plants belong to other animal and plant groups, principally . . . to the true fungi.

The German botanist Robert Hartig, who believed that the acidity of the plant cell and its cellulose walls precluded bacterial infection, took an even stronger stand against the theory that bacteria cause disease in plants:

> They [bacteria] have nothing whatsoever to do with disease processes of plants, in fact I have never on any occasion found schizomycetes [bacteria] in the interior of an intact plant tissue. . . .

Hartig's ideas were familiar to medical bacteriologists of the day who likewise assumed that animals were more susceptible than plants to bacterial infections. Hence it is not surprising that many textbooks of plant pathology in the 1880s and 1890s ignored the question of bacterial diseases of plants or treated it inadequately.

Despite these impediments, evidence was slowly gathering throughout the final decades of the century for the bacterial causation of plant disease. The first recorded observation of the association of a specific bacterium with a plant disease was made by the American plant pathologist Thomas Burrill in a study of pear blight in 1877. Although he failed to isolate the bacterium, it is evident from his conclusion in a later article (1880) on the same disease that he recognized the implications of his discovery:

> The evidence now given of disease *in plants* produced by bacteria contributes something to the germ theory of disease *in animals* and may lead to very important scientific and practical results.

Burrill's suggestion for the bacterial causation of a plant disease appears to have attracted little attention from his contemporaries, possibly because he published in a relatively obscure journal and did not report his results in full. Burrill's work on pear blight was subsequently confirmed (1885) by J.C. Arthur who isolated the causal bacterium in pure culture.

In the years following Burrill's discovery, other botanists began to use bacteriological methods and to suggest bacterial etiologies for certain plant diseases. For example, Adolf Mayer and Martinus Beijerinck in Holland applied bacteriological principles to the investigation in the mid-1880s of tobacco mosaic disease.[3] Despite limited endorsement of the bacterial theory in respect to certain plant diseases, many botanists and bacteriologists refused throughout the 1880s to believe that bacteria were an important cause of plant infections.

This situation began to change in the next decade largely through the efforts of one man, the American plant pathologist, Erwin Frink Smith, who was the outstanding protagonist in the nineteenth century of the bacterial causation of plant disease. During the 1890s he applied Koch's methods to determine the bacteria responsible for several vegetable rots and blights, setting an example of care and precision in plant pathology research. Realizing that the obstacles to the development of plant bacteriology were technical as well as theoretical, he pointed out that investigations of bacterial diseases of plants were frequently performed by bacteriologists without specific training in bacteriological techniques. He recommended that they adopt the high standards of research of medical bacteriology.

In 1895 Smith engaged the German bacteriologist, Alfred Fischer, in a debate which ultimately helped to establish the role of bacteria in plant pathology. The cause of the controversy was Fischer's opposition to the bacterial causation of plant infections, specifically his belief that bacteria were present in diseased plants only as "accidental invaders." The debate reached the pages of the *Centralblatt für Bakteriologie* in the form of three papers by Smith and one by Fischer. Because of the eminence of the journal, these could hardly fail to attract the attention of bacteriologists. Smith, as spokesman for plant bacteriologists, maintained that "there are probably as many plant diseases due to bacteria as there are animal diseases caused by these organisms." (1899) In support of this statement, he presented conclusive evidence of the bacterial causation of eight plant diseases. Partially because of Smith's efforts, bacteria by the turn of the century had come to be widely accepted as plant pathogens.

Pathogenic Protozoa

Protozoa had been observed microscopically since the seventeenth century, but their pathogenicity was not accepted until the late nineteenth. The following is an outline of the work leading to acceptance of their pathogenicity and the effect it had on bacteriology and on the interpretation of research on diseases which today are known to be viral.

According to Dobell, the first observation on record of pathogenic protozoa was made by the Dutch microscopist Antony van Leeuwenhoek who in 1674 in a letter to the Royal Society of London described "oval corpuscles" in the bile of a rabbit. These he had seen with the aid of one of his homemade, single-lensed microscopes about whose use he was so secretive. In the centuries following Leeuwenhoek's report, protozoa were observed on many occasions in the tissues of vertebrates and invertebrates, but no conclusive evidence for their pathogenicity was presented. However, by the 1880s, some investigators were suggesting that malaria, dysentery and the animal disease of surra were caused by protozoa.

These suggestions for the pathogenic activity of protozoa elicited little response from bacteriologists of the 1880s. They had little incentive at a time of discovery and rapid expansion in bacteriology to take active interest in micro-organisms whose pathogenicity, particularly in regard to higher animals, had not been conclusively demonstrated. Nonetheless, as early as 1881 Koch was prompted by research on diseases in which amoebae or spirochaetes had been observed to caution against automatically assuming every infectious agent to be a bacterium:

> It is certainly a one-sided view, although an opinion commonly adopted at the present time, that all as yet unknown infectious materials must be bacteria. Why should not other micro-organisms be just as capable of living parasitically in the animal body? I will not assert that they would only be amoeboid organisms. Other members of the Protistan kingdom[4] are also suspected.

However, as late as 1886, ten years after Koch's confirmation of the germ theory, no specific protozoon had yet been generally recognized as the cause of a major human disease. (Foster, 1965) When attention at last shifted to the protozoa, the process of proving a protozoal etiology was complicated by the fact that most pathogenic protozoa have complex life histories, sometimes with an intermediate host or vector, which were incompletely known at this time. The problem of developing artificial media appropriate for their growth and the absence of an accurate system of classification were further impediments.

As these obstacles were overcome in the 1890s, bacteriologists began to heed the results of recent research on the protozoa.

The work at this time on malaria, Texas fever and nagana, the latter a disease of cattle and horses, removed all reasonable doubt about the pathogenicity of some protozoa for vertebrates. Emanuel Klein's argument in 1891 that infectious diseases were not necessarily bacterial in origin was based on evidence concerning pathogenic protozoa and fungi, and on the negative results of research on diseases such as rabies, smallpox, yellow fever and measles (all viral diseases). Frequent references to pathogenic protozoa in bacteriological literature of the 1890s indicate that they had been widely accepted as disease agents. For example, the American bacteriologist, William Henry Welch, stated in 1894 that

> the pathogenic micro-organisms which we now know to be concerned in causing infections belong to the classes bacteria, fungi and protozoa. . . . (p. 24)

Some investigators looked to the protozoa as a possible cause of the infectious diseases whose etiologies in the 1890s still remained an enigma. Koch in an address to the International Medical Congress of 1890 named thirteen infectious diseases, including smallpox, vaccinia, rabies, influenza, trachoma, yellow fever and cattle plague (all presently known to be caused by viruses), for which "bacteriology has left us completely in the lurch." He directed attention to pathogenic protozoa for a possible solution, at the same time noting the technical problems involved in research on these micro-organisms:

> I am inclined to think that in the case of the diseases referred to it is not at all a question of bacteria, but rather of organized [cellular] generators of disease which belong to completely different groups of micro-organisms. This opinion is all the more warranted by the fact that peculiar parasites belonging to the lowest order of the animal kingdom, the protozoa, have recently been found in the blood of many animals as well as in the blood of humans suffering from malaria.

He went on to say:

> If this problem [of the artificial cultivation of protozoa] is solved, as to which there is no doubt, then in all likelihood, through investigation of pathogenic protozoa and related micro-organisms, a stimulus will be given to bacteriological research which hopefully will lead to the complete disappearance of the mystery still surrounding the etiology of the diseases in question. (p. 756)

A compelling piece of evidence for the impact of protozoal research on bacteriology of the 1890s is the fact that specific structures observed within cells of individuals afflicted with certain infectious diseases were frequently interpreted at this time as stages in the life cycle of protozoa. These structures are today called viral inclusion bodies, found intracellularly in association with viral diseases such as molluscum contagiosum, rabies, and the pox diseases. Consisting of virus particles surrounded by cellular material, they reach dimensions visible in the light microscope. The inclusions, whose true nature was unknown in the nineteenth century, had been observed in connection with certain human, animal and insect diseases earlier in the century. For example, William Henderson and Robert Paterson independently published papers in 1841 on molluscum contagiosum, containing what is said to be the first recorded observations of viral inclusion bodies. These and similar findings were largely ignored until the 1890s.

Inclusion bodies began to attract notice from microbiologists after the appearance in 1893 of Guiseppe Guarnieri's paper describing microscopic structures in the cytoplasm of cells in the lesions of smallpox and vaccinia. He interpreted them to be developmental stages in the life cycle of protozoa which he believed were the specific agents of these diseases. He classified the "protozoa" amongst the pathogenic sporozoa, giving them the names *Cytoryctes variolae* and *Cytoryctes vaccinae*.

Guarnieri's protozoal interpretation of the etiology of smallpox and vaccinia was extended in the early twentieth century to other diseases now known to be viral in which epithelial changes are apparent. For example, some microbiologists believed that cattle plague and herpes zoster, as well as the pox diseases, were caused by protozoa which evoked the vesicular eruptions by intracellular infection. Attempts to elucidate the life history of the protozoal "parasites" by discovering forms which were comparable to the developmental stages of the malarial plasmodium were inconclusive.

By the turn of the century the pathogenicity of protozoa was fairly well accepted. It was now obvious that bacteria were not the only micro-organisms causing disease in humans, and some investigators claimed that protozoa were also the agents of diseases which had long puzzled bacteriologists—diseases which we now recognize as being viral. Thus research in medical protozoology broadened the concept of the nature of the infectious agent and raised the possibility of the existence of still other types of pathogens. But it also led to an interpretation of the nature of inclusion bodies, namely that they were protozoa, which was later found to be erroneous.

The Role of the Host in Infectious Disease

Investigators of the late nineteenth century were not only taking a wider view of the germ theory of disease but they were also coming to recognize the role of the infected individual in the pathogenic process. This had been neglected in bacteriology of the 1880s as investigators in their enthusiasm for the germ theory searched for bacteria which they treated as the sole factor in the manifestation of infectious diseases. However, by the final decade of the century many bacteriologists were recognizing that the discovery and description of pathogenic bacteria did not in itself elucidate the nature of the disease process. As Welch stated in 1894:

> The mere demonstration of a specific micro-organism as the cause of an infectious disease is not a solution of all the etiological problems belonging to the disease. There are many other etiological problems of importance, some relating to the infectious micro-organism and others to the individual exposed to infection. (p. 32)

Under the influence of cellular pathology and recent work on immunization and mechanisms of immunity, bacteriologists began to realize that the etiological agent was only one aspect of the pathogenesis of an infectious disease, and once more to recognize the physiological responses of the body as important factors in the process of infection.

Cellular Pathology. Cellular pathology, a theory of disease which revolutionized mid-nineteenth century pathology, was expounded in 1855 by Rudolf Virchow, the influential German pathologist, anthropologist and politician, and more fully in 1858 in his *Die Cellularpathologie*. To Virchow the cell was the fundamental physiological and morphological unit of the organism, and disease the disturbance of its normal physiological processes. This cell-orientated interpretation of disease temporarily lost favor in pathology after the confirmation of the germ theory when pathogenic bacteria became the focus of attention.

Virchow's view of disease as a cell-based physiological process did not initially incline him to the new bacteriology which emphasized the etiological aspects of infection. He insisted that the constitution of the host played a dominant role in the manifestation of disease. Furthermore, as an adamant proponent of the sociological approach to medicine, he was convinced that hygiene, proper nutrition and adequate housing were important to health. He continued despite bacteriological advances to regard social conditions as prime determinants of the outbreak of epidemics.

Nevertheless, Virchow was not as completely adverse to the bacterial theory as is sometimes supposed. He eventually accepted the evidence for bacterial causation of many infectious diseases and on several public occasions acknowledged bacteriology's contributions to pathology. His well-publicized antipathy to Koch's tubercle bacillus as the cause of tuberculosis was not as extreme as sometimes suggested. He did not totally reject the bacillus as a cause of tuberculosis, but had difficulty in accepting that the various forms of this disease had an identical cause.

Virchow attempted to reconcile cellular pathology and the germ theory by interpreting pathogenic bacteria as external irritants which caused disease in the host through the action on the cells of their metabolic products. These metabolites deranged the normal functions of the cell which thereby retained a primary role in pathogenesis. Recognizing the complexity of the disease process, he disagreed with bacteriologists who assumed that the identification of causal bacteria was tantamount to elucidating the disease itself. He repeatedly pointed out that the cause of a disease and the disease itself were distinct entities. For example, in 1885 he maintained, referring to tuberculosis, that

> knowledge of the bacillus, no matter how important it is for the full understanding of the origin of a disease process, still in no way explains the disease process itself, nor does it obviate the special investigation of it.

Virchow also pointed out the fallacy of attributing

> every impurity to bacteria on the sole ground of its contagiousness. It may be said that a contagious disease affords suspicions of a bacterial origin, but it should not be called simply bacterial. To do so hinders further research and lulls the conscience to sleep. (1898)

Thus Virchow's view of infectious disease as a phenomenon involving the interplay of cellular processes, pathogenic micro-organisms and sociological factors was later found to be a more accurate assessment than the narrow bacteriological approach epitomized in Koch's postulates.

Immunization and Immunity. Bacterial virulence was known in the 1880s to be a highly variable quality which was related to such things as the species of bacterium, the number of bacteria in the infecting dose and their portal of entry into the body. More importantly, the virulence of a given bacterial species was not fixed but varied with exposure, whether natural or experimental, to various physical and biological conditions. "Virulence," as Emile Duclaux put it, "is in perpetual development." (1886, p. 179)

The physiological aspect of immunity, namely the body's mechanism of defense, was not extensively explored until the late nineteenth century. It had been noticed centuries earlier that certain diseases did not recur in the same individual, but no adequate explanation for the phenomenon was known. The ancient and dangerous procedure of variolation, first practiced in the Orient and introduced into Europe in the early eighteenth century, was an initial attempt to provide protection against smallpox by the inoculation of material from the mild form of the disease (variola minor). Sporadic attempts were made in the eighteenth and early nineteenth centuries to discover artificial means of producing immunity to other infectious diseases. The only notable success was Edward Jenner's smallpox vaccine which he developed empirically in total ignorance of the identity of the infectious agent and of the manner in which the vaccine protected inoculated individuals from subsequent attacks of the disease.

The modern phase of immune prophylaxis began in 1880 with Pasteur's announcement of a living attenuated vaccine against chicken cholera. He had discovered by chance that the injection of an old culture of the causal bacteria prevented a recurrence of the disease in the same chicken. As a tribute to Jenner, he chose to call this substance a "vaccine." Grasping the principle of microbic variation, Pasteur went on to the empirical development between 1880 and 1888 of vaccines against anthrax, swine erysipelas and rabies by attenuating the specific infectious agents through exposure to oxygen and/or heat. He also found that "passage," the transfer of infectious material from one animal to another in series, heightened or in some cases diminished the virulence of the micro-organism.

Until the mid-1880s the physiological and biochemical reactions of the organism to infectious disease tended to be overlooked in bacteriological investigations. The preoccupation of bacteriologists with the isolation and identification of pathogenic bacteria left little opportunity for the exploration of pathophysiology.[5] Although the interaction between micro-organism and host was implicit in the empirical research on immunization by Pasteur and his associates, the practical goal of developing vaccines was given priority over inquiry into the body's susceptibility or resistance to infection. The few theories of immunity proposed before 1884 were based on the principle of bacterial activity rather than on the responses of the host.

The role of the body in infectious disease first received widespread attention in bacteriology in 1884 when the Russian zoologist, Ilia Metchnikoff, advanced his phagocytic theory of immunity. He believed that protection was due to phagocytic cells in the blood and tissues which destroyed bacteria and foreign particles.

Metchnikoff's cellular theory of immunity was opposed by the humoral theory introduced in 1888 by the American biologist and hygienist, George Nuttall, and later amplified by others. These humoralists held that the main defence of the body against micro-organisms lay in chemical substances in the body fluids. Support for this viewpoint came from contemporaneous work on toxins, antitoxins and antisera.

By the mid-1890s it was apparent that both cellular and biochemical mechanisms were involved in the immune reaction. A few years later Paul Ehrlich, the eminent bacteriologist and immunologist, advanced the sidechain theory of immunity which was an attempt to explain biochemically the complex manner in which the body protects itself by producing antibodies against antigens.

Efforts such as these reflected the concern of microbiologists in the 1890s with the physiological aspects of disease. The German bacteriologist and immunologist, Hans Buchner, commented in 1891 on this reorientation of research:

> One finally began to notice that with the discovery of a disease agent, with the popular formulation that now 'the cause' of each and every disease was found, the actual scientific issue was still in no way exhausted. On the contrary, questions about the how and why proliferated more than ever on all sides, and with these starting points research naturally turned back to the pathological-

physiological area from which it had originally taken its departure. . . . Little by little things moved towards a deeper understanding, towards a physiological theory of the process of infection and of its reverse, which one designates as immunity or recovery reactions.

The Recognition of Infectious Diseases of Unknown Etiology

As bacteriology developed into a sophisticated science with a complex methodology and technology, bacteriologists became increasingly suspicious of purportedly bacterial etiologies based on incompletely substantiated information. For example, André-Victor Cornil and Victor Babes, the authors of a popular textbook on bacteria, decided that the bacterial etiologies which some investigators suggested for the "eruptive fevers" (infectious diseases accompanied by a skin eruption, such as smallpox and measles), had not been adequately confirmed by experiment. In their opinion, the causes of these diseases remained open to further investigation.

With the rejection of research which did not conform to the bacteriological standards of the late nineteenth century, the existence of these enigmatical infectious diseases (for which even careful research had failed to provide an experimentally verifiable explanation) became increasingly evident. Methods and techniques for identifying bacteria could not be faulted since they had been used with great success in discovering the cause of a succession of bacterial diseases. Yet the fact remained that bacteriologists had failed to find the cause of many important infectious diseases. Welch admitted in 1894 that

> . . . we have a large number of infectious diseases which have thus far resisted all efforts to discover their specific infectious agents. Here belong yellow fever, typhus fever, dengue, mumps, rabies, Oriental pest, whooping cough, smallpox and other exanthematous fevers, syphilis, and some other infectious diseases in human beings. It will be noted that many of these are the most typically contagious diseases, which it might have been supposed would be the first to unlock their secrets. . . . (p. 31)

The German bacteriologist Ferdinand Hueppe also commented on current ignorance of the cause of acute exanthematic diseases, adding:

> Negative findings are conspicuous in precisely those contagious diseases whose infectiousness is greatest and are perhaps explainable by the fact that bacteria are not involved.

Plant pathologists as well as medical bacteriologists were aware of infectious diseases with unknown etiologies. For example, Erwin Smith who demonstrated in 1888 that peach yellows (later identified as a viral disease) was infectious, was still deliberating six years later over the cause of the infection:

> There has been much speculation concerning the nature of this disease, inasmuch as climate and soil do not seem to originate a plainly communicable malady, and no fungi, bacteria, or animal parasites have been identified as the cause. No fungus has been found associated with it constantly, and it is almost certainly not a bacterial disease, statements to the contrary resting upon evidence no careful mycologist or bacteriologist would for a moment be willing to accept.

Summary

By the final years of the nineteenth century, bacteriology had evolved from a science devoted largely to the discovery and description of pathogenic bacteria into one in which these interests were supplemented by research on the physiological, biochemical and sociological aspects of infectious disease. Over the previous few decades bacteriology had acquired an effective methodology, an array of experimental techniques and a record of solid achievement in elucidating the etiological and some of the pathophysiological and biochemical aspects of infectious disease. Furthermore, by the 1890s it had become specialized into medical, veterinary and plant bacteriology, and had also given rise to the science of immunology. These fields of inquiry, as well as protozoology and aspects of mycology, were components of the new science of microbiology.

The 1880s and 1890s had also seen the appearance of institutes, journals, textbooks and courses of instruction devoted to the subject of bacteriology or microbiology. The two most important microbiological institutes of the late nineteenth century were the Pasteur Institute, founded in Paris by world-wide subscription in 1888, and the Institut für Infektionskrankheiten in Berlin, established by the German government in 1891 with Koch as director. One of the first textbooks to give substantial treatment to pathogenic bacteria was A. Magnin's *Les Bactéries.* (1878) After 1885, journals specializing in the publication of research on pathogenic bacteria began to appear.

At the Sorbonne Emile Duclaux in 1879 began to teach the first course in microbiology ever given. For several years he delivered a series of lectures on fermentation and medical bacteriology. These lectures were the basis of his book, *Ferments et Maladies.* (1882) In 1884 Koch offered a comprehensive course in medical bacteriology at the University of Berlin, and Emile Roux began instruction in microbiology in 1888 upon the opening of the Pasteur Institute. Edgar Crookshank in 1886 offered a course in clinical bacteriology at King's College, London, where he also founded the first bacteriological laboratory in Britain. Bacteriology began to be taught in American medical schools and agricultural colleges in the middle and late 1880s.

Thus by the late nineteenth century microbiology had acquired the appurtenances of a mature science devoted to the investigation of all aspects of microbial life. Because of its scope, microbiology attracted investigators from various sciences, each with a different approach to microbial research, thus providing an environment appropriate for the growth of new ideas. This fertile environment was conducive to the development of new concepts and experimental approaches.

Notes

1. By the late nineteenth century microscopic fungi were recognized as a major cause of infectious diseases in plants. But these microorganisms were thought, rightly as we now know, to play a relatively minor role in infections of animals and man.
2. For an account of the late nineteenth century attitude of botanists, pathologists and bacteriologists to the extension of the bacterial theory of disease to plants, see Smith, 1911, vol. 2, pp. 9–22.
3. This research is discussed in Chapter 4.
4. The Protista comprise the bacteria, protozoa, and unicellular algae and fungi.
5. Kohler maintains that in the 1880s the challenge of discovering pathogenic micro-organisms was great enough to defer questions about the physicochemical aspects of disease.

3

The Infectious Agent: Exceptions to the Conventional View

With acceptance of the germ theory, most investigators assumed that agents of infectious disease were micro-organisms which could be detected with the light microscope, retained by bacterial filters, and cultured on artificial media. Koch's postulates were only of use if the infectious agent under investigation actually possessed these properties. Nonetheless, speculations concerning infectious agents which did not have all these properties were occasionally made. Although these speculations did not lead directly to the concept of the virus, they raised the possibility that infectious agents might possess properties differing from those of known micro-organisms. Some of the suggestions for the existence of infectious entities which were either submicroscopic, filterable, or incapable of being grown *in vitro* will be considered in turn below.

More extreme views were held by certain scientists who still believed, despite advances in bacteriology, that infectious diseases were not caused by micro-organisms nor by any other form of living particle. For example, Charles Creighton, the British pathologist and medical historian, maintained up to his death in 1927 that miasmata, soil poisons, and seismic disturbances were factors in the rise of epidemics. This view colors his *History of Epidemics in Britain* whose two volumes appeared in 1891 and 1894. Creighton refused to accept bacteria as causal agents of disease, although he did not deny that they were sometimes concomitants of it. Those diseases which did not arise from miasmata or some other condition in the external environment, he regarded as resulting from a persistent physiological disturbance which somehow developed autonomy as a disease. As a result of these unorthodox ideas and, additionally, his denunciation of Jennerian vaccination, he found himself relegated to the fringes of medical science from the late nineteenth century onwards. His contemporaries either ignored his work completely or regarded it as blatantly outmoded.

The neurologist Henry Charlton Bastian also held highly

controversial views about the causation of infectious disease. He subscribed to a theory of zymotic disease which incorporated elements of his belief in heterogenesis, the generation of micro-organisms from living or nonliving matter. An antagonist of the germ theory of fermentation and disease, he opposed in the 1870s such eminent scientists as Pasteur, Koch and Tyndall. It has been said that his attacks on adversaries helped indirectly to strengthen the position of bacteriologists by pointing out weaknesses in their experimental techniques which were then rectified.

Even amongst those who adhered to a theory of animate contagion, there was disagreement about the fundamental nature of the disease agent. For example, Lionel Beale and Antoine Béchamp believed that infectious diseases were caused by minute living particles derived from the tissues of higher organisms. Their ideas about the etiology of infectious disease and the nature of the infectious agent will be discussed since it is pertinent to consider whether, as theories of particulate infection, they influenced the development of the concept of the infectious agent in the late nineteenth century.

Submicroscopic Organisms

It was commonly assumed by scientists engaged in microbial research in the 1860s and early 1870s that microbial forms existed which were too small to be detected with the microscope. The morphology and manner of reproduction of micro-organisms were incompletely known. Some people held that microscopically visible bacteria developed from submicroscopic precursors or "germs" somewhat in the manner of animals and plants. This idea was difficult to disprove because of the limited resolving powers of contemporary microscopes and the crudeness of bacterial staining techniques.

The first experimental evidence appearing to support speculations about submicroscopic germs was provided by John Tyndall in his well-known experiments on "floating matter in the air." In studies which provided an argument against the spontaneous generation of micro-organisms, he used a specially designed chamber and a concentrated beam of light which revealed particles in the air, particles "beyond the reach of the microscope." He subsequently found that by directing the beam at tubes of nutrient media he could see particles by diffracted light which he could not detect with the microscope. In 1876 Tyndall explained the significance of this discovery:

> 'Potential germs' and 'hypothetical germs' have been spoken of with scorn, because the evidence of the microscope as to their existence was not forthcoming. Sagacious writers had drawn from their experiments the perfectly legitimate inference that in many cases the germs exist, though the microscope fails to reveal them. Such inferences, however, have been treated as the pure work of the imagination, resting, it was alleged, on no real basis of fact. But in the concentrated beam we possess what is virtually a new instrument, exceeding the microscope indefinitely in power. Directing it upon media which refuse to give the coarser instrument any information as to what they hold in suspension, these media declare themselves to be crowded with particles—not hypothetical, not potential, but actual and myriadfold in number—showing the microscopist that there is a world beyond even his range.

Tyndall's work called attention to the submicroscopic bacterial "germ" and to the inadequacy of the contemporary microscope for microbial research.

By the late 1870s the microscope had been greatly improved by the addition of the achromatic lens, by now a common component, and the substage condenser. The latter was first introduced in 1873 by the German physicist and optician, Ernst Abbé, and further improved in 1888 by an achromatic version. Staining and histological techniques for the observation of bacteria had also advanced, largely through Koch's contributions. In 1877 he introduced a method of fixing and staining bacteria on cover glasses and also described the use of photography in microscopical work. One consequence of the improved view of the microbial world now available to investigators was to divert attention from the question of submicroscopic organisms. But, before long, they were again being considered in bacteriology. The impetus came from two sources.

First, the results of earlier work describing the finite resolving powers of the microscope had now become known to many microbiologists. Abbé had reported in 1873 that the resolving powers of the microscope, hitherto thought to be limited only by technical inadequacies, were restricted to objects which were at least as large as the wavelength of light. It was for this reason that investigators came to recognize the possibility of the existence of micro-organisms which forever would be invisible in the light microscope because they were smaller than the wavelength of light. Emile Duclaux, writing in 1886, suspected that some infectious liquids which were "perfectly limpid in appearance" contained micro-organisms which "our microscopes, however improved in their present construction we suppose them to be, will always be unable to demonstrate to us." (p. 28) Technical improvements reached such a level that by 1890 the microscope in skilled hands was capable of resolving structures at the theoretical limits of visibility (about 0.25 micron).

The second reason for revived interest in submicroscopic organisms was the discovery that causal agents could not always be detected microscopically in infectious materials. This did not mean that use of the microscope was abandoned. On the contrary, microscopy continued to be an integral part of research on infectious diseases. Even Pasteur, a scientist known for his resourcefulness, searched with the microscope for at least four years for visible "microbes" of rabies (a viral disease). Indeed, on two occasions he suspected that he had seen them—in 1881 in the form of encapsulated bacteria and in 1883 as fine particles in the brain tissue of rabid animals. In 1885, having abandoned the microscopical aspect of his research on rabies, Pasteur made the remarkable discovery that a vaccine could be prepared without prior identification of the infectious agent. More will be said below about this achievement.

Filterable Micro-organisms

Although filtration as a basic procedure in chemistry dates from antiquity, only in the latter half of the nineteenth century did it come to be used for the separation of micro-organisms from the fluids in which they are contained. Casimir Davaine was one of the first to employ the bacterial filter in research on an infectious disease. In his investigation of anthrax in the 1860s, he used the placenta of a living guinea pig and also a "porous partition" (cloison poreuse)[1] to separate the bacilli from blood in an attempt to prove that they, rather than a bacterial toxin, produced the disease. In 1871 Ernst Tiegel, a Swiss physiologist, improved the procedure of bacterial filtration by connecting unglazed clay cells to a Bunsen air pump. By this method he also succeeded in isolating anthrax bacilli from blood. Pasteur used filters of plaster and later porous porcelain vases in his bacteriological work of the late 1870s.

In the early 1880s Charles Chamberland and Emile Roux increased the filtering surface by employing a hollow cylinder for bacterial filtration. The first cylinder was the stem of a clay pipe which they sealed at one end with wax. In 1884 Chamberland presented a perfected model of unglazed porcelain which could be attached to a water tap. (See Fig. 3) The Chamberland filter became a common fixture of bacteriological laboratories and was used domestically for the filtration of drinking water. Another filter widely employed in bacteriological work was the Berkefeld filter of "Kieselguhr" or diatomaceous earth. It was introduced by H. Nordtmeyer in 1891 and named after the owner of the mine in which the filtering properties of the earth were first noted. In the same year Shibasaburo Kitasato introduced the prototype of a filter which was later improved and sold commercially under his name. Many types of filters, differing primarily in composition and pore size, were subsequently developed. All have pores small enough to hold back the majority of bacteria.

An understanding of the factors involved in the deceptively simple procedure of filtration did not keep pace with improvements in the device itself. For many years it was assumed that filters acted as mechanical sieves which retained particles but allowed dissolved substances to pass through. However, by the final years of the nineteenth century microbiologists had made two observations about bacterial filtration which conflicted with the sieve idea. First, micro-organisms were not invariably held back by bacterial filters. This had become apparent from observation of their use for the purification of drinking water; all filters after a period of time were permeable to a certain number of micro-organisms. Also repeatedly demonstrated was the fact that under conditions suited to bacterial growth, some micro-organisms subsequently appeared as contaminants of the filtrate, regardless of the type of filter employed.

Second, it was known that certain soluble substances were retained by filters. For example, Pasteur learned as early as 1883 that chemicals such as strychnine, even though dissolved, could be adsorbed by filters. Taking this information into account in his research on anthrax, he decided that the noninfectious filtrate obtained by filtering infected blood was not conclusive evidence that the infectious agent was a bacterium rather than a soluble toxin. In short, it had been realized that the permeability or nonpermeability of a bacterial filter to a substance was not always a reliable guide to its particulate size.

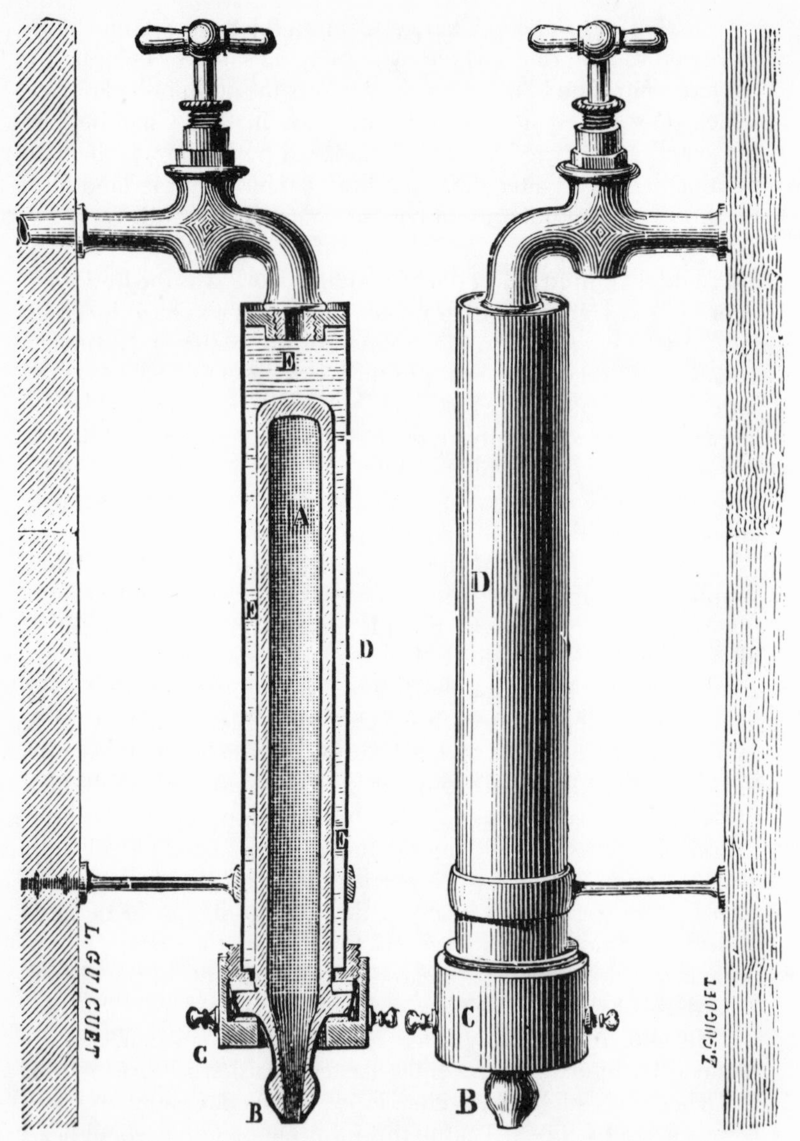

Fig. 61. — Filtre Chamberland, coupe et élévation.

Fig. 3 Chamberland Filter.

Infectious Agents: Problems of *In Vitro* Cultivation

Most microbiologists of the late nineteenth century assumed agents of infectious disease to have the potential for *in vitro* growth and, like Pasteur (1880), entertained the hope of eventually obtaining "artificial cultures of all viruses [infectious agents]." However, it was known from experience that the appropriate conditions for the artificial cultivation of certain micro-organisms were difficult to attain. Consequently, it was more logical for them to attribute an infectious agent's failure to grow in artificial cultures to the choice of inappropriate nutrient media rather than to suspect that it might be an obligate parasite incapable of *in vitro* development. For example, Ernst Levy and Felix Klemperer in 1898 attempted to explain the failure to culture the causal agent of smallpox (a viral disease) by suggesting that perhaps it was not a bacterium and therefore made "different demands on nutritive substrates than those bacteria whose cultivation has hitherto been successful." However, they appear not to have explored this possibility further.

Perhaps the most outstanding example of research on a disease in which the causal agent could not be grown *in vitro* was Pasteur's investigation of rabies in the 1880s. He found that the infectious agent could not be detected microscopically nor grown in plate cultures, although he did not deny the possibility of one day succeeding in these endeavors. Yet despite ignorance of the nature of the infectious agent and the failure of traditional bacteriological methods, Pasteur achieved the goal of his research on rabies, the development of a vaccine.

His method for obtaining pure *in vivo* cultures of the agent consisted of introducing infected material directly into the brains of dogs by trepination, neural tissue having been found to be specific for its growth. The agent was then passed through rabbits by successive inoculations until it had reached stability at maximum virulence; Pasteur had thereby obtained what he called a "fixed virus." He considered the spinal cords of inoculated rabbits to contain pure cultures of the causal agent at maximum virulence. He could attenuate these cultures by degrees to produce vaccines of decreasing virulence. The administration of these vaccines first to animals and then to humans in a series of injections, beginning with the fully attenuated agent and proceed-

Fig. 4 A cartoon from *Vanity Fair* of Pasteur with rabbits used in his research on rabies. (From the original cartoon in the Wellcome Museum, by courtesy of the trustees.)

ing to the most virulent, gradually augmented resistance to the disease. Pasteur's achievement received worldwide acclaim, attracting people to his laboratory for what became known as the "Pasteur treatment" for rabies. To the scientific community he had demonstrated that infectious agents whose identity was unknown and which had failed to develop in nonliving media could be grown in receptive tissues of experimental animals and attenuated to form vaccines. Pasteur's appreciation (1884) of the importance of this achievement for future research on infectious disease is evident in the following:

> For a long time in the future the art of preventing diseases will be closely associated with virulent diseases in which the microbes elude our investigations. Therefore it is a point of major scientific importance that the protective inoculation against a virulent disease can be discovered, if really necessary, without having the actual virus [infectious agent] at one's disposal and while remaining ignorant of the isolation and cultivation of the corresponding microbe.

A suspicion that some infectious agents might be unable to grow in artificial media began to develop towards the end of the century. The use of Koch's culture methods had not led to success in the cultivation of the agents of a number of infectious diseases. Recollecting the characteristic of obligate parasitism in plants and animals, some microbiologists thought it probable that certain micro-organisms might also be dependent on a living host.

For example, the French bacteriologist Saturnin Arloing suggested in 1891 that the agents of diseases acquired only by direct contact with an infected individual, such as rabies, syphilis and vaccinia, were probably obligate parasites and hence restricted in their development to living tissues; conversely, other pathogenic micro-organisms, such as those of typhoid, cholera and anthrax, alternated saprophytic and parasitic modes of existence.

William Henry Welch, writing in 1894, classified bacteria either as parasites or saprophytes, concluding that the great majority of bacteria were obligate saprophytes which grew on or in nonliving materials. Those capable of existing both inside and outside the host he called either facultative parasites or saprophytes, according to where they spent most of their existence. The remainder he described as obligate parasites which "we suppose cannot find, or at least only exceptionally can find, natural conditions suitable for growth outside of the living body. . . ." (p. 13) In this latter class Welch included certain bacteria, such as the bacillus of tuberculosis, which could be cultivated by "special methods" in artificial media. But even for obligate parasites he had not completely excluded the possibility of *in vitro* growth.

For the remainder of the century most microbiologists continued to assume that the inability to culture certain infectious agents *in vitro* was a technical problem which would be overcome in the future. Thus there was still no full recognition of the fact that some agents of infectious disease were completely incapable of growth on or in nonliving media, that is that they were absolute parasites.

Lionel Beale's Theory of Infectious Disease

Lionel Smith Beale was an English physician who spent the major part of his career at King's College, London. There he taught anatomy and physiology and later became professor of pathological anatomy (1869–1876) and the principles and practice of medicine (1876–1896). Two things brought him to prominence in the 1860s: first, his pioneer work on the use of the microscope in medicine and the development of histological techniques, and second, his unique theory of biological organization which arose from his microscopic studies. His private course in the application of the microscope in clinical medicine, reputedly the first ever given, was the basis for his popular textbook on the microscope.

Beale's views on infectious disease were a facet of his theory of biological organization. This, briefly stated, was that living structures are composed of "germinal matter," roughly comparable to protoplasm, and "formed material," such as the cell wall and intercellular substance, which he believed were derived from germinal matter but no longer alive. As early as 1864 Beale suggested that infectious diseases probably were not due to "the propagation and transference of vegetable organisms [microorganisms], but to small particles of living animal matter which have descended from the germinal matter of one organism and have been transferred to another." In his report of 1866 to Her Majesty's Cattle Plague Commissioners, he attributed the cause of this disease (now known to be a virus) to minute particles derived from germinal matter. These particles, which could be less than

1/100,000th of an inch in diameter (about 0.25 micron), were small enough to pass into the tissues and blood where they reproduced.

Beale's theory of infectious disease reached full development in two books published in 1870. He maintained that "disease germs" were direct descendants of living "bioplasm," a word which he coined at about this time as a substitute for "germinal matter." The process began with a physiological change which caused rapid multiplication in the blood of bioplasmic particles. These became progressively degraded with each new generation. After several generations of "retrograde evolution," microscopic "disease germs" specific for each disease were produced which might be transmitted to other organisms. Here they entered the blood where they multiplied, producing changes peculiar to each disease. Beale was an ardent opponent of the theory of bacterial causation of disease, insisting that bacteria present in infected tissues were either the result or "accidental concomitants" of disease.

Beale's ideas about infectious disease appear to have evoked little comment during the last decades of the century, doubtless because the germ theory was well established by this time. Furthermore, he was no longer actively engaged in scientific research, his interests by this time having turned to philosophical and religious speculation. At the time of his death in 1906 his scientific reputation rested primarily on his microscopic work of the 1860s, his bioplasmic theory having been all but forgotten.

Antoine Béchamp's Theory of Infectious Disease

Antoine Béchamp, a French chemist who held doctorates in pharmacy and medicine, was professor of chemistry and pharmacy at Montpellier from 1857 to 1874. In the latter year he became dean of the Free Faculty of Medicine in Lille. He, like his contemporary Beale, continued long after the confirmation of the germ theory to assert that infectious diseases were caused by living, microscopic particles which differed in origin and composition from those of the known micro-organisms.

Béchamp's formulation of the theory of "microzymas" began in the course of his work on fermentation in the 1860s. The concept of the microzyma first arose during microscopical observations of chalk in which he described organisms "smaller than all those we know, smaller than all Infusoria or Microphytes that we study in fermentations; . . . they are alive and mature, although without doubt very old. They act with singular energy as ferments. . . ." (1866) He gradually extended to all living phenomena the functions of microzymas, originally confined to fermentations. By the time of the publication in 1883 of Béchamp's book, *Les Microzymas*, they were defined as the basic vital and indestructible elements composing living beings and had become the core of a new doctrine concerning "organization and life."

An important aspect of the doctrine was Béchamp's interpretation of infectious disease. He maintained that the tissues of all living things contained microzymas which, when their normal physiological activities were disturbed, were able to evolve into bacteria. The process of transformation consisted of a gradual fusion of microzymal particles, resulting finally in the formation of "bacteria." Bacteria thus evolved were able to infect other beings where they were presumed to provoke a "dyscrasia" or abnormal state of the tissues, leading eventually to the formation of more bacteria. Béchamp rejected the conventional bacterial theory of disease, asserting that there was no such thing as an inherently pathogenic micro-organism since it was merely a transitory state of the microzyma. In his opinion "the primary cause of our disease is in us, always in us. . . ." (1883) Thus according to Béchamp, bacteria were not the cause but rather the result of disease.

With the development of bacteriology in the late nineteenth century, it became evident that the microzymal explanation of infectious disease was in disagreement with an ever-increasing body of evidence. The fact that it was founded mainly on speculation rather than on accurate observation and experimentation deprived it of scientific credibility. For example, Pasteur in 1886 decried the microzyma as a "thing of pure fantasy" and Cornil remarked in the same year that Béchamp had provided "no direct proof of his doctrine of the microzymas."

Aside from the fact that the bioplasmic and microzymal theories of disease conflicted with the fundamental premise of the germ theory, there is evidence that Beale and Béchamp were estranged from orthodox science. Towards the end of the century Beale adopted an extreme form of vitalism which had been implicit in his earlier disagreements with Thomas Huxley and other "materialists." Beale subsequently fully explained his vitalistic theory in a series of speculative papers in the *Lancet*.[2] Judging from the letters published in the same journal in response to these articles, Beale's opinions, at least on this subject, were poorly received. More importantly, by now it was obvious that his theory of biological organization and his ideas about the causation of infectious diseases conflicted with the cell doctrine and with the germ theory of disease.

Béchamp's antagonism towards Pasteur and his followers had grown increasingly bitter since their initial disagreement in the 1860s over priorities in research on fermentation. The conflicts with Pasteur, which sometimes interrupted proceedings in the Academy of Medicine in Paris, intensified Béchamp's resistance to ideas which jeopardized the microzymal theory. By the end of the century, Béchamp had become almost totally alienated from the scientific community, finding support from only a handful of students.

It would appear that both Beale and Béchamp had to a great extent lost scientific repute in the eyes of their contemporaries. Hence it may be assumed that their interpretations of infectious disease, including their views on the nature of the infectious agent, had little or no influence on pathological theory of this period.

Notes

1. In 1882 it was demonstrated that bacteria often cross the placental barrier. See Théodoridès, pp. 79 (footnote), 85 and 117.
2. The substance of Beale's many papers in the *Lancet* on "Vitality" may be found more conveniently in his book, *Vitality: An Appeal, an Apology and a Challenge*.

4

The Discovery of the Filterable Infectious Agent

The tobacco mosaic virus, which causes a mottling of the tobacco leaf, has always figured prominently in research on viruses. In 1892 Dimitri Iosifovich Ivanovski, a graduate student in botany at the University of St. Petersburg, noted that the causal agent of tobacco mosaic disease passed through a bacterial filter. Six years later in 1898, Martinus Willem Beijerinck,[1] unaware of Ivanovski's work, also found that the agent was filterable and described it as a contagium vivum fluidum. Both men have been credited with the discovery of the virus. Their contributions to the development of the concept of the virus and the work of Adolf Mayer, who was the first to do microbiological research on tobacco mosaic disease, will be considered in this chapter.[2]

Early Research on Tobacco Mosaic Disease

Adolf Eduard Maydolf Mayer. Adolf Mayer, a German chemical technologist, began to study tobacco mosaic disease in 1879 at the request of a Dutch agricultural society concerned with the damage done to tobacco crops in the provinces of Gelderland and Utrecht. The Society sent samples of infected tobacco leaves to Mayer, the director of the Agricultural School in Wageningen from 1876 to 1904. In 1883 Mayer decided that the disease was infectious because he could transmit it from one plant to another by inoculations of sap extracts from the leaves of diseased plants. This is reputed to be the first experimental transmission of a viral disease of plants. (Hanon, p. 37) He named it "die Mosaikkrankheit" (the mosaic disease) because of the mosaic of dark and light spots present on infected leaves.

One of the first to employ Koch's methods to plant bacteriology, Mayer made repeated attempts to isolate causal microorganisms in artificial cultures. When these efforts failed, he filtered sap extracted from diseased plants through a single layer of filter paper and determined that the filtrate was still infectious. He concluded that microscopic fungi could not be the cause of the disease because they were sufficiently large to be retained by the paper. However, he found that the filtrate became noninfectious after filtration through two layers of paper. Since Mayer believed that a noncellular substance would have passed through both layers, he concluded that the causal agent could not be "an enzyme-like body." (p. 44) He found, furthermore, that the infectivity of the sap was destroyed by heating it at 80° for several hours. He interpreted these observations to mean that the causal agent was "organized," i.e. cellular, and hence either a microscopic fungus or a bacterium. He had already eliminated fungi as being too large to pass through filter paper and he also thought (falsely as Ivanovski soon demonstrated) that a double layer of filter paper was sufficient to retain bacteria. For these reasons he concluded:

> The mosaic disease of tobacco is a bacterial disease, of which, however, the infectious forms are not isolated nor are their form and mode of life known. (pp. 44–45)

Dimitri Iosifovitch Ivanovski. Mayer's work was known to a young Russian student by the name of Ivanovski who began his career in microbiology and plant physiology with an investigation of tobacco diseases, including tobacco mosaic. While a student at the University of St. Petersburg, he spent three summers (1887–1889) doing research on tobacco plantations in Bessarabia, the Ukraine and the Crimea. This research was the basis for a paper which Ivanovski and a fellow student published in 1890. They asserted that Mayer's Mosaikkrankheit was really two distinct diseases, neither of which they believed to be infectious. Unable to detect micro-organisms in infected plants and failing to transmit the disease by inoculation, they concluded that "parasites" could not be the cause. Instead they attributed the disease to disturbances in the normal life processes of the plants.

In 1892 Ivanovski published the two papers upon which his reputation as the "discoverer" of the filterability of viruses is based. He now admitted that tobacco mosaic was an infectious disease for he had found that sap from infected plants, contrary to Mayer's observation, retained its infectious properties after filtration through two layers of filter paper.

Ivanovski suspected that a bacterial toxin might be the cause of tobacco mosaic disease. This idea was inspired by the discovery in 1888 of the diphtheria toxin by Emile Roux and

Fig. 5 Dimitri Iosifovich Ivanovski. (Photograph courtesy of V.M. Zhdanov of the D.I. Ivanovski Institute of Virology, Moscow.)

Fig. 6 Ivanovski, the bearded figure to the left of center, with one of his classes. (Taken from a pamphlet in Russian by G.R. Matukhin entitled *D.I. Ivanovski: Centenary of His Birth*, University of Rostov, 1964.)

Alexandre Yersin. Ivanovski set out to test his hypothesis by passing infected sap through a bacteria-proof Chamberland filter. He found that after the filtrate had been inoculated into twelve healthy plants, symptoms of the disease appeared in nine: The filtrate was obviously infectious. This result apparently did not surprise Ivanovski for it appeared to support his idea of a bacterial toxin which, being a soluble substance, would pass through the Chamberland filter. Hence he simply noted that "if the mosaic disease is due to bacteria, it might be expected that the sap filtered in this way would have infectious qualities." (1892a, translation, p. 149)

The obvious problem was to isolate the toxin-producing bacteria. Despite experiments requiring "a great deal of time and trouble," his attempts to grow the "tobacco microbe" on artificial substrates were unsuccessful. "Under certain special conditions" (which Ivanovski did not describe), he had been able to see "the growth of the tobacco microbe and then to determine its presence in the tissues of infected plants." (1892a, translation, p. 151) This encouraged him to believe that artificial cultivation of the pathogen would eventually be achieved.

In his short report of 1892 to the Academy of Sciences in St. Petersburg, Ivanovski emphasized by means of italics that "*the sap of leaves infected with tobacco mosaic disease retains its infectious properties even after filtration through Chamberland filter candles.*" (1892, p. 69) He then suggested that a bacterial toxin was the most likely cause of the disease:

> According to current views, the infectiousness of the filtrate can be explained most simply by assuming the presence of a poison elaborated by bacteria present in the tobacco plant and dissolved in the filtered sap.

Another "equally acceptable explanation," he continued, was that "the bacteria of the tobacco plant passed through the pores of the Chamberland filter candles, even though I checked the filter . . . before each use and convinced myself of the absence of fine cracks and openings." (1892, pp. 69–70)[3] Evidently recognizing the inconclusive nature of his research on tobacco mosaic disease, Ivanovski closed by expressing his hope to conduct further experiments "to clarify these questions." Despite these intentions, Ivanovski turned to research on the fermentation of alcohol and published nothing more on tobacco mosaic disease until after the appearance in 1899 of Beijerinck's fourth paper on this subject.

Martinus Willem Beijerinck. Beijerinck's initial research on tobacco mosaic disease, like Ivanovski's, was influenced by Adolf Mayer with whom he collaborated at the Agricultural School in Wageningen where Beijerinck taught from 1876 to 1884. At Mayer's suggestion, Beijerinck made microscopical studies of infected tobacco plants and attributed his failure to detect causal bacteria to his then inadequate training in microbiological techniques. Encouraged by his isolation in 1887 of the root-nodule bacterium of leguminous plants (*Bacillus radicicola*), Beijerinck made a second attempt to find a causal micro-organism, this time looking specifically for small toxin-producing bacteria either in diseased plants or in soil around their roots. Unsuccessful in these endeavors, he asserted that "tobacco mosaic disease is an infectious disease but it is not provoked by microbes." (1900, p. 297)[4]

He resumed his investigation in 1897 with the advantage of the facilities of the new microbiological laboratory at the Polytechnical School in Delft where Beijerinck had been appointed professor of general microbiology in 1895. Unaware of Ivanovski's prior discovery, Beijerinck reported that "sap from infected plants, after having transversed a porcelain filter . . . remains virulent." (1900, p. 297) After searching unsuccessfully for tiny anaerobic bacteria which might have passed through the filter, he decided that the filtrate was bacteriologically sterile. Nonetheless, he found that the mosaic disease could be transmitted successively through an "unlimited" number of plants.[5] For this reason he concluded that the pathogen had the capacity to multiply within living plants even though he considered its filterability indicative of a noncellular substance.

The properties of the agent of tobacco mosaic disease were paradoxical: Its ability to infect and to multiply suggested a micro-organism; its filterability and resistance to alcohol, weak formalin and desiccation were usually characteristics of nonliving material. Beijerinck had not been able to detect the infectious agent with the microscope and he was unwilling to decide about its physical characteristics solely on the basis of its filterability. He knew from previous experiments that allegedly bacteria-proof filters did not invariably retain particles of bacterial dimensions nor did they invariably filter soluble material. In addition, he was aware that some anaerobic bacteria possess extremely small, filterable spores. However, he discounted spores as a possible cause of tobacco mosaic disease when he found that infectivity was completely destroyed by a single exposure to a temperature of ninety degrees centrigrade.[6]

Deciding that traditional bacteriological techniques did not serve his purpose, Beijerinck devised an experiment which he hoped would determine conclusively whether the infectious agent was "corpuscular" (cellular or particulate) or "dissolved" (noncellular and nonparticulate). He placed sap from infected tobacco leaves on the surface of a solidified agar medium. After allowing ten days for diffusion to occur, he cleaned the surface and then discarded a thin layer of agar. The fresh agar thus exposed was removed and injected into tobacco plants which subsequently developed tobacco mosaic disease. Beijerinck explained this experiment as follows:

> A virus [infectious agent] composed of small, discrete particles will remain on the surface because it cannot diffuse into the molecular pores of the agar plates. In this situation the deeper layers of agar will not become virulent. Conversely, a water-soluble virus [infectious agent] will be able to penetrate a certain depth into the agar plates. (1899, p. 29)

In Beijerinck's opinion the results of his experiment could be interpreted in only one way:

> Although diffusion occurred over a distance of only several millimeters, nonetheless it seems to show in actual fact that the virus [infectious agent] is really liquid or dissolved and not corpuscular. (1900, p. 298)

This conclusion was based on the assumption that only soluble materials are able to diffuse; cells and particles, which do not form homogeneous or 'true' solutions, would therefore be incapable of diffusion.[7] Hence Beijerinck concluded that the disease agent, because it appeared to have the ability to diffuse, had to be regarded as "liquid" or "dissolved" and therefore noncellular.

The concept of a soluble substance which induced changes in the host plant was not entirely new to Beijerinck. In papers of 1888 and 1897 on plant galls, he had postulated the existence of "gall-forming substances" which circulated in growing tissues of the host plant where they provoked changes at the cellular level resulting in the production of galls. Beijerinck drew attention in his publications on tobacco mosaic disease to the similarity between the gall-forming materials and the agent of tobacco mosaic disease in respect to their solubility and capacity to

Fig. 7 Martinus Willem Beijerinck at the age of 70.

diffuse in the meristem, the undifferentiated tissue of plants where active cell division occurs.

Beijerinck was convinced that the pathogen was not a chemical substance since its ability to reproduce indicated to him that it was alive. For this reason and the fact that the agent was soluble, he concluded that the cause of tobacco mosaic disease was a living infectious agent in fluid form for which he coined the term contagium vivum fluidum.

Beijerinck's contagium vivum fluidum represented a radically new concept of an infectious agent. Since Rudolf Virchow's dictum "e cellula ex cellula" of 1858, it had been assumed that anything which reproduced was alive and cellular in form. By proposing the existence of a living fluid substance, Beijerinck had broken the conceptual linkage between life, as evidenced by reproduction, and the cell.

Despite Beijerinck's certainty of the agent's capacity to multiply, he could not discover any sign of its replication outside the plant, either in artificial cultures or in the filtered sap. Furthermore, he observed that only growing portions of the plant were subject to infection. Therefore he decided that the pathogen multiplied exclusively in tissues which were undergoing cell division. He had also encountered this "important property" in the gall-forming substances which, like the contagium vivum fluidum, were active only in dividing tissues. (1899, pp. 29–30)

The fact that reproduction occurred only if the infectious agent was "bound to the living protoplasm of the host plant" (1899, p. 30) was associated, Beijerinck believed, with its dissolved state. He argued that even a microscopically invisible and dissolved pathogen, if it could be grown *in vitro*, would produce detectable color and refractive changes in nutritive media. These changes did not occur when it was placed on substrates suited for the growth of plant bacteria. Moreover, Beijerinck thought that this manner of active molecular replication under *in vitro* conditions, while not inconceivable, was difficult to imagine:

> It is not easy to accept a process of division in the molecules which would lead to their multiplication, and the concept of 'self-supporting molecules' [sich ernährenden Molekülen], which consequently must be assumed, seems unclear to me, if not contrary to nature. (1899, p. 31)

Thus, reluctant to accept the idea of a living and actively self-reproducing molecule, Beijerinck suggested a way in which replication might occur in a passive fashion:

> . . . it is to some extent an explanation to consider that the contagium in order to reproduce itself must be incorporated into the living protoplasm of the cell into whose multiplication it is so to speak passively included. (1899, p. 31)[8]

However, he was also uncertain about this suggestion for he admitted that "the incorporation of a virus [infectious agent] into living protoplasm, even if factually established, cannot be regarded as a completely comprehensible occurrence." (1899, p. 31)[9] However, this general mode of replication reminded Beijerinck in many ways of that of amyloplasts and chromoplasts "which likewise only grow along with the cellular protoplasm but nonetheless lead an independent existence and function on their own." (1900, p. 301)

The Controversy between Ivanovski and Beijerinck

The controversy between Ivanovski and Beijerinck over the nature of the agent of tobacco mosaic disease began in 1899 after the appearance in the *Centralblatt für Bakteriologie* of the latter's fourth paper on the disease. (1899) Ivanovski, who apparently had not seen Beijerinck's previous papers which had appeared exclusively in Dutch journals, claimed priority for the discovery of the filterability of the agent. Ivanovski went on to say that he found Beijerinck's agar experiment "open to even more criticism than my filtration experiment" because "we understand the process of filtration through porcelain candles better than the occurrences which take place by bringing agar and fluid plant sap together." (1899, p. 251)[10]

Turning to his own as yet unpublished experiments, Ivanovski announced that he had transmitted the mosaic disease successively from plant to plant by means of the filtered sap and concluded, as Beijerinck had done previously, that "the virus [infectious agent] multiplies in the living plant." (1899, p. 252) It is significant that Ivanovski made no reference in this paper to his earlier suggestion that the cause of tobacco mosaic disease was a bacterial toxin. In a comment written later in life, Ivanovski explained his reasons for abandoning the toxin idea:

> I concluded my short preliminary report of 1892 with the supposition that the mosaic disease is a bacterial one and that either the bacterium itself is present in the filtrate from the Chamberland filter, i.e. that it can pass through the filter due to its insignificant size; or perhaps that a soluble toxin . . . is capable of causing the entire spectrum of the disease. I could not at that time decide which of these suppositions was most likely as an insufficient number of experiments was available concerning the serial inoculation of the filtered (sterilized) sap for several consecutive generations. Later I carried out such investigations and showed that plants which had become diseased from the inoculation of sterilized sap are capable of passing on the infection to another healthy plant, and this to a third, etc. Consequently, the hypothesis regarding the toxin dissolved in the sap lost ground, and the hypothesis that the microbe of tobacco mosaic disease passes through the pores of an unglazed ceramic filter remained; my investigations stopped at this point. I only returned to them six years later, in 1898.[11]

In other words, Ivanovski realized that the pathogen could not be a toxin if it had the ability to reproduce and thereupon rejected this idea.

Beijerinck immediately responded to Ivanovski's paper of 1899 with a short article in which he "willingly acknowledged" Ivanovski's priority for the discovery of the infectiousness of the filtrate, explaining that at the time of writing he had been unacquainted with Ivanovski's work on tobacco mosaic disease. (1899a)[12] Beijerinck repeated the agar experiment with a viral filtrate derived from tobacco which had been grown under bacteria-free conditions. In this way he sought to eliminate the possibility that microbes with the capacity to grow within agar had created a path along which the disease agent had migrated without true diffusion. The results seemed to him to indicate once again that the agent was able to diffuse, thus reassuring Beijerinck that he had been correct in describing it as a contagium vivum fluidum. By 1900 he was occupied with other research and never returned to the investigation of tobacco mosaic disease.

The culmination of Ivanovski's research on the mosaic disease was his doctoral dissertation of 1902 which was republished in similar form in the *Zeitscrift für Pflanzenkrankheiten* in 1903. Ivanovski again criticized Beijerinck's agar experiment as an inadequate basis for asserting that the infectious agent was fluid and for the first time offered experimental evidence for his opinion. He demonstrated that ink particles, under the same conditions as the tobacco sap, moved from the surface of an agar plate into its interior. Rejecting for this reason Beijerinck's contagium vivum fluidum, he decided that "the most likely conclusion is that the contagium is contained in the sap in the form of solid particles." (p. 25)[13]

Ivanovski cited several additional observations to support his argument that the mosaic disease was caused by a particulate agent, but realized that it was necessary to isolate a causal micro-organism in order to prove his point conclusively. Resuming his microscopical investigations of infected plants, he searched for micro-organisms small enough to pass through bacterial filters. In fixed and stained cells he found tiny amoeba-like structures which he called "zooglea."[14] (pp. 34–35; also see Fig. 8, numbers 9–16, in this book) He thought they might be the causal agents of tobacco mosaic disease.

He also encountered "crystalline deposits" in the cytoplasm of infected cells. (See Fig. 8, numbers 3, 4 and 6, and numbers 10, 11 and 12) They had the form of thin, refractive plates which developed striations in acid media. Ivanovski decided that the crystalline plates were not the etiological agents but rather that they represented "a reaction of the cells to the irritation produced by the parasites [of tobacco mosaic disease]." (p. 35)[15]

Although unable to isolate the "zooglea," Ivanovski nonetheless proposed that the etiological agent might be a spore-forming micro-organism. These spores, he asserted, rather than the micro-organism itself, were filterable. He believed that this hypothesis would account for the infectiousness of the filtrate and, if the spores were able to germinate "only in the living plant or generally only under optimal conditions," that it would also explain the failure of attempts to culture the "microbe" *in vitro* from the infectious filtrate. (p. 37) He regarded the resistance of the pathogen to heat and desiccation to be further indication that spores might be present in filtrates of infected sap. Ivanovski then attempted to isolate from unfiltered infected sap the vegetative form of the bacterium supposedly producing the spores. From this sap he claimed to have grown bacteria which produced tobacco mosaic disease when inoculated into tobacco plants but his attempts to isolate and identify a causal micro-organism were inconclusive.[16] Ivanovski in effect admitted his failure by concluding the paper with the following statement:

> On the whole the problem of the artificial cultivation of the microbe of the mosaic disease must be reserved for future investigations. (p. 41)

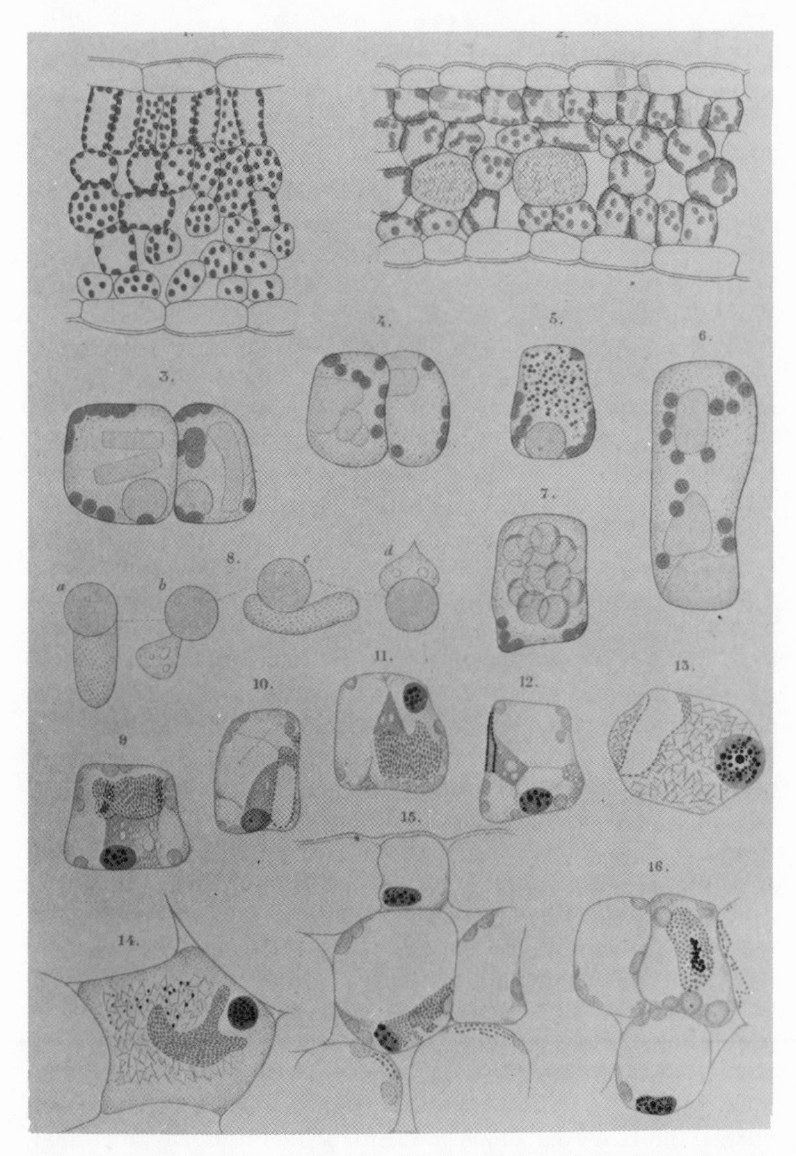

Fig. 8 From Ivanovski's "Uber die Mosaikkrankheit der Tabakspflanze," Z. Pflkrankh, 1903, 13: 1-41. The "zooglea" are visible as oval or round bodies in numbers 9-16. The "crystalline deposits" (tobacco mosaic virus in crystalline form) may be seen as irregularly shaped plates in numbers 3, 4 and 6, which depict unfixed and unstained preparations fixed in alcohol and stained with methyl blue and eosin.

The Agent of Tobacco Mosaic Disease: Assessment of the Theories of Ivanovski and Beijerinck

Ivanovski. Ivanovski attempted to explain the unconventional properties of the agent of tobacco mosaic disease within the context of the germ theory by attributing the cause of the disease first to toxin-producing bacteria and then to a spore-forming micro-organism. Working within this strict bacteriological framework, he did not interpret his own experimental findings as indicating the existence of a unique type of infectious agent. While he can certainly be credited as the first to recognize the filterability of the causal agent of an infectious disease, he cannot justly be termed the "discoverer" of the virus, as some have claimed,[17] nor to have formulated an original concept of it. Ivanovski's research on tobacco mosaic disease—his initial discovery of the agent's filterability and his hypotheses concerning its bacterial nature—does not appear to have had an effect on the development of the concept of the virus. His papers of 1892 had been published only in Russian journals and his work was virtually unknown in the West before 1899, a year after filterable agents were described by Beijerinck and others.

Ivanovski's work, even after it began to appear in German journals, seems to have evoked little comment from contemporary scientists. The reasons for this indifference can only be surmised. One possible explanation is that he provided no conclusive evidence to support his suggestion that bacteria cause tobacco mosaic disease. Another possibility is that he did not publish any original contributions to research on tobacco mosaic disease between 1892 and the completion of the dissertation of 1902. In his three short papers published around the turn of the century (1899, 1901, 1902), he was primarily concerned to establish his priority for the discovery of the filterability of the infectious agent. By the time his dissertation had been published (1902), the two conflicting concepts of the virus, the microbial and the nonmicrobial, had been formulated for several years and were to remain in competition throughout the early decades of the twentieth century.

Beijerinck. Beijerinck, like Ivanovski, began his investigation of tobacco mosaic disease with the premise that it was a bacterial infection, and for years employed bacteriological methods in attempts to identify the causal agent. However, he eventually became convinced that it was neither microscopically visible nor cultivable in artificial media. Having in effect abandoned Koch's postulates, he began to consider the infectious agent from a standpoint outside the bacteriological tradition. Ultimately he formulated a concept of the causal agent which for the first time clearly differentiated it from pathogenic micro-organisms. He presented an entirely new view of the infectious agent; his contagium vivum fluidum was a microscopically invisible and filterable substance which was alive and multiplied only within living cells.

This interpretation anticipated in several aspects the modern concept of the virus, namely the fact that it is an infectious, reproducing entity which is an obligate parasite of living cells. Beijerinck gave his concept wide applicability by suggesting that it was "extremely probable that many other plant diseases whose origin is unknown, but which are not caused by parasites, will be attributed to a fluid contagium." (1900, p. 310)

Hence it would appear that Beijerinck rather than Ivanovski should be credited with the formulation of a concept which agrees to some extent with the modern definition of the virus. Particularly remarkable is Beijerinck's suggestion that the agent reproduced by passive inclusion in the protoplasm of dividing cells. This property of obligate intracellular replication, rather than submicroscopic size or filterability, is now recognized to be one of the crucial differences between viruses and micro-organisms.

It was the very originality of Beijerinck's concept of the virus which made its acceptance difficult around the turn of the century. His idea of the contagium vivum fluidum appeared to be contradicted by two doctrines of late nineteenth century science: the germ theory of infectious disease and the cell theory.

For two decades the germ theory had guided the ideas and the experimental methodology for research on infectious diseases. Consequently, microbiologists were reluctant to accept Beijerinck's view which ascribed to certain infectious agents properties differing fundamentally from those of pathogenic micro-organisms. For example, Emile Roux, a bacteriologist and pupil of Pasteur, admitted in 1903 that the concept of a contagium vivum fluidum was "very original"; nonetheless he suggested that the possibility of a tiny spore-bearing micro-organism should be considered before accepting Beijerinck's idea. (p. 11)

Beijerinck's view of a living, reproducing, noncellular material[18] was also difficult to reconcile with the theory promulgated by Virchow that the cell is the fundamental unit of all living things. For in proposing the concept of a contagium vivum fluidum, Beijerinck had broken free from the chain of reasoning which linked reproduction with the cell.

The implications of the concept of a contagium vivum fluidum did not escape the notice of the German biochemist and enzymologist, Carl Oppenheimer. To him Beijerinck's "liquid contagia" suggested the possibility of vital forces in "dissolved cells," that is, in noncellular substances. He suggested that certain diseases of humans, such as scarlet fever, smallpox and syphilis, might also be caused by a contagium vivum fluidum. If this were true, he thought "our concept of the 'cell' may undergo such a radical change as to admit that vital forces can actually act in unorganized [noncellular] media."

Others such as the German bacteriologist, Ernst Joest, refused to consider modification of the cell theory, stating that there was no such thing as a substance which was both living and soluble. He maintained that "a contagium vivum fluidum can only exist in the form of organized cell individuals [micro-organisms]." Hence Beijerinck's contagium vivum fluidum was not only "unthinkable" as the cause of tobacco mosaic disease but also as the cause of other infectious diseases. (pp. 412–413)

Thus scientists accepting the cell as the basic unit of life and the germ theory as the explanation of the cause of infectious diseases were frequently reluctant to endorse the concept of the contagium vivum fluidum. In a tradition where the cell theory was dominant, it was not easy to accept the idea that a noncellular substance could reproduce. Furthermore, Beijerinck's concept of the infectious agent as a living, multiplying and yet noncellular substance raised considerable experimental as well as theoretical difficulties. The technology and experimental methodology of the period were not developed sufficiently for the investigation of entities which were later found to have dimensions between those of simple molecules and cells. Nonetheless, the nonmicrobial concept of what we now know as the virus remained for many decades a viable, albeit subordinate, component of research on infectious diseases, representing an alternative to the microbial concept.

Notes

1. Biographies of Ivanovski and Beijerinck are found in Appendix B.
2. A paper by the author entitled "Early concepts of the virus: the work of D.I. Ivanovski and M.W. Beijerinck" has been accepted by the *Bulletin of the History of Medicine.*
3. Ivanovski tested his filters for cracks by trying to force air through them while they were submerged in water. He also considered the absence of bacterial growth in nutrient media, which had been filtered several months previously, to be further evidence that his filters were faultless. (1892, p. 70)
4. See the Bibliography for a list of papers and translations of Beijerinck's research on tobacco mosaic disease.
5. Having inoculated only one series of tobacco plants, Ivanovski could not have observed that the disease was serially transmissable and hence that the causal agent multiplied in the plants.
6. Spores are only destroyed by repeated exposures to temperatures close to 100 degrees centrigrade.
7. The concept of the 'true' solution was based at this time on the theories of Jacobus van't Hoff and Svante Arrhenius. In 1885 van't Hoff proposed that substances in dilute solution exist as molecules governed by the same laws as gas molecules. This theory did not hold for aqueous solutions of acids, bases and salts. These exceptions were explained by Arrhenius in his theory of ionization (1887) in which he suggested that a dissolved electrolyte immediately dissociates to produce ions. It was also believed that only 'soluble' substances, hence small molecules or ions, were able to diffuse. (Leicester)
8. The wording of the comparable sentence in the paper of 1900 is somewhat different: "The virus [infectious agent], without the power to grow by itself, is drawn into the growth of the dividing cells, and multiplies there enormously without losing any of its individuality." (p. 300)
9. The mechanism of viral replication was not explained until the 1950s. See Chapter 6.
10. Ivanovski's dissatisfaction over the agar experiment has turned out to be warranted. Beijerinck's opinion about the physical form of the contagium vivum fluidum was founded on the mistaken assumption that capacity for diffusion is an accurate criterion for differentiating dissolved substances from particles and cells. It is now known that there is no precise difference in the state of dispersion of large particles and small organisms. Moreover, the concept of the macromolecule and accurate methods for determining the size of large

molecules were developments of the twentieth century. See Chapter 6, note 3. Hence at the turn of the century neither Ivanovski nor Beijerinck could have known that molecules, macromolecules and cells form a continuous dimensional gradient.

11. Quoted by G.M. Vaindrakh in his article in Russian, "D.I. Ivanovski. Biographical sketch" in *On Two Diseases of Tobacco. Tobacco Mosaic Disease*, Moscow: Medgiz, 1949, translated by J.M. Irvine, pp. 50–51 of the unpublished translation.
12. Like most scientists outside Russia, Beijerinck was unfamiliar with Ivanovski's work until it was brought to his attention by the latter's paper of 1899. Ivanovski's previous work had appeared only in Russian journals although his papers of 1890 and 1892 had been reviewed respectively in the *Botanisches Centralblatt* (1891, 47: 370–373) and in the *Beihefte* of the same journal (1893, 3: 266–268).
13. Judged from present knowledge, Ivanovski, like Beijerinck, erred in using the agar matrix as a guide to the physical form of the disease agent.
14. Zooglea are jelly-like masses of micro-organisms.
15. It is now believed that Ivanovski was in fact describing a virus in crystalline form. Crystals (actually two-dimensional 'paracrystals') of tobacco mosaic virus sometimes form in infected cells and have been shown with the electron microscope to be composed largely of the virus and a volatile substance which is probably water.
16. Ivanovski's proclaimed success in transmitting the disease by means of bacterial cultures may have resulted from contamination of the cultures with the mosaic virus.
17. For example, Y.I. Milenushkin and V.I. Basalkevich, "The discoverer of the world of viruses. Centenary of the birth of Dimitri Iosifovich Ivanovski (1864–1920)," *Vop. Virus* (1964, 9: 521–526). (In Russian)
18. Beijerinck was not the first to conceive of a living unit smaller than the cell. As Hall has pointed out, various biologists during the latter half of the nineteenth century sought a living entity or "metastructure" which lay between the simple molecule and the cell in size and composition. Similar ideas were expressed in medicine. As discussed in Chapter 3, both Beale and Béchamp ascribed the cause of infectious diseases to living, subcellular entities, respectively the bioplast and the microzyma.

5

The Development of the Microbial and Nonmicrobial Concepts of the Virus to 1900

Research on viral diseases from 1898 to approximately the middle of the twentieth century was marked by two conflicting concepts of the virus. One, the microbial, defined the virus as a minute micro-organism, usually submicroscopic and filterable; the other, the nonmicrobial, defined the virus as an infectious noncellular substance. This chapter traces the development of these two concepts in the course of various investigations of viral diseases conducted between 1898 and 1900.

The Microbial Concept of the Virus

Friedrich Johannes Loeffler and Paul Frosch. In 1897 Loeffler and Frosch became chairmen of a commission established by the German government to recommend means to control foot and mouth disease which was threatening the country's cattle industry. Both men were bacteriologists trained by Koch. Loeffler was a professor at the veterinary school at Greifswald and Frosch was professor of bacteriology at Koch's Institute of Infectious Diseases in Berlin.

They found that standard bacteriological techniques were ineffectual in research on foot and mouth disease; causal micro-organisms were not detected in extensive microscopical studies of lymph taken from the vesicles present in the mouth and on the hooves of infected animals, nor could they be grown in artificial cultures. However, this lymph produced foot and mouth disease when injected into young cattle.

In March 1898 Loeffler and Frosch published three reports in which the filterability of an animal virus was described for the first time. The fact that it was possible to immunize animals against the disease with mixtures of immune blood and vesicular lymph prompted Loeffler and Frosch to test it for the presence of a dissolved substance with immunizing activity. This they did by passing it through bacterial filters. To check the effectiveness of their filters, they added a culture of *Bacillus fluorescens* to diluted lymph which was then filtered two or three times. They were convinced that the filtrate was bacteria-free when these readily detectable bacteria failed to appear in cultures of the filtrate.

To their surprise, they found on injecting small amounts of the filtered lymph into calves that foot and mouth disease appeared after the normal incubation period with the usual symptoms. The experiments were repeated many times on cows and pigs; all acquired the disease in its typical form. "We had the impression," Loeffler and Frosch reported, "that the infectiousness of the lymph had not been influenced by filtration." (p. 389) As far as they knew the finding was without precedent for they were not acquainted with Ivanovski's work on tobacco mosaic disease, and Beijerinck's first paper on the contagium vivum fluidum had not yet been published. In their opinion there were two possible explanations for this observation:

> Either the bacteria-free filtered lymph contained a dissolved poison of extraordinary power or the as yet undiscovered agents of the infectious disease were so small that they were able to pass through the pores of a filter definitely capable of retaining the smallest known bacteria. (p. 389)

Considering the first possibility, Loeffler and Frosch turned to Ludwig Brieger's calculations for the strength of the tetanus toxin. On the basis of these figures they estimated the comparative toxicity of the agent of foot and mouth disease after passage through two experimental animals. They decided that a toxin in order to remain virulent after successive dilutions in the bodies of two animals would have to possess such potency as to be "simply unbelievable." Consequently, they concluded that the infectiousness of the filtrate was not due to a dissolved substance but rather to the activity of a living agent with the ability to multiply in animals. These agents, they continued, "must be so small that they pass through the pores of a filter which is guaranteed to retain the smallest bacteria." (p. 391)

The "influenza bacillus" with a length of 0.5 to 1.0 micron was the smallest bacterium then known. Loeffler and Frosch calculated that if the agent of foot and mouth disease were only one tenth or one fifth as large, it would be invisible in the microscope even with the most up-to-date oil immersion systems. The presence of pathogenic organisms of submicroscopic size, they decided, would provide a simple explanation for their failure to detect causal agents in the infectious lymph.

Pending the outcome of further research, they suggested that "the agents of numerous other infectious diseases, such as smallpox, cowpox, scarlet fever, measles, typhus, cattle plague, etc., which up to now have been sought in vain, might also belong to this group of minute organisms." (p. 391)

In a paper published a few months later, Loeffler reported that foot and mouth disease had been transmitted through a series of six animals. The fact that the last animal acquired the disease as quickly as the first, even though he calculated that it had received only two-billionths of the original amount of lymph, confirmed the belief that the infectious agent multiplied in the animals and hence was alive. Loeffler also found that diluted lymph lost virulence after repeated filtration through a fine-pored Kitasato filter.[1] This further convinced him that the causal agent of foot and mouth disease was not soluble but "corpuscular" or cellular.

John McFadyean, the English bacteriologist, wrote in 1908, that Loeffler and Frosch's discovery of the filterable agent of foot and mouth disease "at once attracted general attention, and gave a great impetus to the investigation of the nature of the virus in those diseases which had hitherto baffled investigation conducted on ordinary bacteriological lines." (p. 68) Thus their concept of the infectious agent and their experimental methods had considerable influence on the development of the concept of the virus and on the methodology of future research on viral diseases.

Their concept held that the agent of foot and mouth disease was a pathogenic organism which was smaller than familiar micro-organisms and therefore microscopically invisible and filterable through all but the finest bacterial filters. Its *in vitro* cultivation depended merely upon discovering the appropriate growth medium. Thus the difference between this pathogen and ordinary micro-organisms was primarily a matter of size and not a question of a completely new type of infectious agent. Their concept of the agent required only minor modifications of current ideas about the nature of disease-bearing micro-organisms.

Loeffler and Frosch's investigation of foot and mouth disease also established the technical basis for research on viral diseases. They hoped, as Loeffler subsequently acknowledged, that "the same experimental methods would be used . . . to ascertain whether similar tiny living organisms might be present in [other] infectious products of disease." (1911, p. 5)

Microscopy had not proved useful in positively identifying the cause of foot and mouth disease, although negative findings had led Loeffler and Frosch to conclude that the agent was submicroscopic. They gave filtration a new interpretation by placing emphasis on what was filtered rather than on what was retained. They also emphasized the use of inoculation experiments on animals whose pathological reactions were the only positive evidence then obtainable for the presence of viruses. They regarded their failure to culture the agent of foot and mouth disease on nonliving media as merely a temporary technical difficulty. They and other investigators persisted after the turn of the century in their efforts to culture viruses *in vitro*.[2]

Loeffler and Frosch thus established an experimental methodology which in the early twentieth century was widely adopted in research on human and animal viral diseases. It was based primarily on three techniques: microscopy and filtration to determine the absence in infectious materials of microscopically visible organisms, and serial animal inoculation experiments to provide evidence for the presence of infectious agents in the filtrates. The weakness of this experimental approach, as investigators began to realize in the twentieth century, was that it provided little positive information about the nature of these pathogens beyond the fact that they were extremely small, infectious, and able to multiply in animals.

Edmond-Isidore-Etienne Nocard and Emile Roux. Almost simultaneously with the report of Loeffler and Frosch on foot and mouth disease, Nocard and Roux with their collaborators at the Pasteur Institute in Paris published the results of an investigation of bovine pleuropneumonia.[3] Nocard, in his presentation of the work to the International Congress for Hygiene and Demography in Madrid in April 1898, ascribed the cause of the disease to "a microbe of an extreme tenuity" which after many fruitless attempts had been cultured under very specific conditions. Small collodion sacs containing a trace of infected pleural exudate and a meat infusion had been surgically inserted into the peritoneal cavity of rabbits and allowed to incubate. After a lapse of fifteen to twenty days, examination of the contents of the sacs using high magnification and strong lighting revealed "an infinity of small, refringent and mobile points of such great tenuity that it is difficult, even after staining, to determine their form exactly." (p. 244) Nocard and Roux decided that the "points" were living organisms which had multiplied to such an extent as to make the medium slightly cloudy. Cows inoculated with subcultures developed symptoms of pleuropneumonia, thereby confirming that these organisms caused the disease. Nocard also reported that on one occasion he had succeeded in culturing the organism under conventional *in vitro* conditions and therefore was assured that it could be grown outside the body. This finding was confirmed in 1900 when the pathogen was again cultivated on solid nutrient substrates.

Nocard made no reference in the paper of 1898 to attempts to filter the agent of pleuropneumonia, but subsequent events indicated that he had considered its filterability. Loeffler, who attended the presentation of Nocard's paper in Madrid, asked at the end of the talk if the causal agent was filterable. Nocard replied negatively but changed his mind in 1899 after discovering that it passed through coarse Berkefeld and Chamberland filters if the lymph containing it was greatly diluted. (Loeffler, 1911, p. 4) Nocard, Roux and their associates had thus discovered another filterable agent of disease. But, unlike that of foot and mouth disease, it could be demonstrated in the microscope and grown in artificial cultures.

An important result of the discovery of the causal agent of pleuropneumonia was to lend support to the microbial concept of the virus. Direct microscopical evidence of a filterable pathogen which could be artificially cultured seemed to substantiate Loeffler and Frosch's idea of a small micro-organism. If this were true of the agents of foot and mouth disease and bovine pleuropneumonia, then perhaps the agents of other hitherto unexplained infectious diseases might also be exceedingly minute, filterable and potentially culturable micro-organisms.

Even though the agent of pleuropneumonia was microscopically detectable, its size (at the limits of visibility) increased the likelihood of the existence of submicroscopic microbial agents of infectious disease. For, as Nocard and Roux pointed out, "it is perfectly permissible to conceive of the existence of still smaller microbes, which, instead of being *on this side* of the limits of visibility, as is the case with the microbe of pleuropneumonia, would be *on the other side* of these limits; in other words, one can admit that microbes invisible to the eyes of man exist." (p. 248) Roux remarked five years later that the organism of pleuropneumonia "forms a connection between ordinary bacteria and those which the microscope is incapable of showing." (p. 10) The agent of pleuropneumonia was regarded as the smallest microscopically visible member of a continuous chain of minute organisms which extended from ones of microscopically visible dimensions to those which were beyond the reach of the microscope. The existence of submicroscopic organisms, which, as discussed in Chapter 3, had long been only a speculation, now appeared to be a virtual certainty.

Nocard and Roux's work also affected the experimental methodology of early research on viral diseases. It seemed reasonable to assume that if the agent of pleuropneumonia were really so closely related to submicroscopic disease agents, then the techniques employed in the investigation of pleuropneumonia could also be applied to research on diseases in which microscopically invisible pathogens were suspected. They remarked in 1898 that the discovery of the agent of pleuropneumonia "gives hope of comparable success in the study of other infectious materials for which the microbe is presently unknown." (p. 248) They also maintained that research on diseases caused by submicroscopic pathogens could be successfully undertaken if the appropriate medium for their *in vitro* growth was discovered. The difficult but ultimately successful cultivation of the agent of pleuropneumonia was promising in this regard. Recognizing that the growth of submicroscopic organisms in nonliving media might remain microscopically undetectable, Nocard and Roux asserted that the animal inoculation experiment was the only sure method for demonstrating the presence of microscopically invisible disease agents.

The investigation of pleuropneumonia demonstrated that the procedure of filtration was not free of interpretative problems. Nocard and Roux found that the pathogen, even in highly dilute suspension, was retained by finer grades of the Berkefeld and Chamberland filters. It was becoming evident that the type and grade of filter and the concentration of the material to be filtered were factors to be considered in determining the filterability of an infectious agent. By 1903 sufficient work with bacterial filters had been done for Roux to state:

> In order to classify a microbe amongst those which traverse filters, it is not sufficient to ascertain that it has traversed the [filter] wall; it is necessary to know under what conditions [it has done so]. (p. 10)

In addition, the discovery that the microscopically visible agent of pleuropneumonia was able to traverse certain types of bacterial filters demonstrated that, contrary to the expectations of Loeffler and Frosch, the attributes of submicroscopic size and filterability were not necessarily connected. In the early twentieth century the property of filterability rather than submicroscopic size was to become the primary indication that a disease agent was a virus. (See Chapter 6) Such was not the case with Sanarelli's discovery of the myxoma virus.

Guiseppe Sanarelli. The International Congress for Hygiene and Demography of 1898 learned of the existence of yet another submicroscopic disease agent when Sanarelli, an Italian bacteriologist who for a time directed the Institute of Hygiene in Montevideo, Uruguay, presented a paper on rabbit myxomatosis. The agent was classified as a virùs on account of its submicroscopic size, even though its filterability was not reported for some years. However, Sanarelli noted in 1898 that centrifugation produced an infectious serum which did not contain micro-organisms.

Although the paper is entitled "The myxoma virus," the subtitle, "Contributions to the study of disease agents beyond the visible," suggests that Sanarelli considered his investigation of a specific pathogen to have general significance. In the first section of the paper he discussed a class of infectious diseases, including rabies and syphilis,[4] in which no causal micro-organisms had been found. Of particular interest is Sanarelli's admonition that the properties of these pathogens could not be understood "unless the greatest part of our knowledge . . . of the nature of the organized viruses [micro-organisms] known at present is given up." "However," he continued, "since it is unlikely that there are unorganized [noncellular] infectious agents in nature,[5] one is indeed compelled to believe that certain diseases are produced by organisms which are so small that they are scarcely visible to the human eye, even when it is aided [by the microscope] (as for example the bacterium of pleuropneumonia demonstrated some time ago by Nocard and Roux), or [by organisms] which are completely invisible. The new disease which I now have to report gives me the opportunity to investigate this matter somewhat more closely." (p. 866)

The stimulus for Sanarelli's research had been an outbreak of myxomatosis in laboratory rabbits at the Institute of Hygiene. Sanarelli suspected that the agent of the disease was not related to any of the familiar pathogenic organisms when his application of every microbiological technique "which can be devised in a laboratory today" (p. 868) failed to reveal an etiological agent. Nonetheless, he found that the disease could be transmitted successively from rabbit to rabbit and that it increased in virulence after passage through several animals.

Sanarelli did not mention his attempts to filter infectious fluids, but his research assistant later reported that when infected and diluted blood had been filtered twice through small-pored Berkefeld filters, the final filtrate was found to be sterile. (Centanni, p. 201) (This sterility was probably due to the use of a fine filter with what we now know to be a relatively large virus. The electron microscope shows the myxoma virus to be approximately 0.29 × 0.23 microns.) He also used "spontaneous coagulation" and centrifugation, by which means he obtained serum which was "optically completely pure and thoroughly sterile [bacteriologically]." (p. 869)[6] The serum proved to be infectious when injected into rabbits.

Skeptical of the existence of "unorganized" infectious agents, Sanarelli adopted the view that the myxoma pathogen was "organized" or cellular, despite his failure to identify it in the microscope. He was convinced that it was related to "no other organized being which we are accustomed to regard at present as the cause of specific disease," (p. 868) but did not describe the precise form such an agent might have.

Cornelis Johan Koning. In 1899 Koning, a Dutch pharmacist interested in the problems of tobacco cultivation, published a paper on tobacco mosaic disease. The paper was based on laboratory research conducted at the University of Amsterdam and on field research at Wageningen and Amerongen in Holland. He had assumed in a previous publication (1898) that the disease was microbial in origin but now reported that he was unable to detect or to culture causal micro-organisms. Koning found, however, that the disease could be transmitted serially from plant to plant by means of the infected sap. As a result he came to the by now familiar conclusion that the infectious agent was able to multiply and that "this multiplication can be ascribed to nothing other than living organisms which temporarily evade observation." (p. 71)

Lacking direct evidence for the existence of a causal micro-organism, Koning was forced to rely on indirect methods to indicate its nature. He found that sap from diseased tobacco plants remained infectious after filtration as well as after brief exposure to a temperature of one hundred degrees centigrade which is capable of destroying most forms of life. For these reasons he suggested that the etiological agent was a filterable micro-organism which probably possessed heat-resistant spores. (p. 74) Although his efforts at *in vitro* cultivation were unsuccessful, he continued to assume that the agent was potentially cultivable under these conditions. He also noted that the pathogen had "some correspondence" (presumably on account of its filterability) with the agent of foot and mouth disease. He supposed the "organism" of tobacco mosaic disease to be larger than that of foot and mouth disease,[7] believing that only the filtrate of the former decreased in virulence with repeated filtration. (p. 75)

Koning's first reference to Ivanovski and Beijerinck appeared in 1900 in his book, *Der Tabak*, a year after their papers on the mosaic disease had been published in the *Centralblatt für Bakteriologie*. In a footnote to the chapter on tobacco mosaic disease he stated that only after publication of his paper of 1898 had he learned of their research. Koning acknowledged Ivanovski's priority for the discovery of the filterability of the agent of tobacco mosaic disease and also observed that Beijerinck had made a similar observation in 1899. He offered no further comment on their work.

Thus apparently unfamiliar with previous work on tobacco mosaic disease, Koning had come to the tentative conclusion that the causal agent was a filterable (though not necessarily submicroscopic), spore-bearing micro-organism with the potential for growth in nonliving media.

The Nonmicrobial Concept of the Virus

Albert Woods. In the late 1890s an increasing number of physiologists, chemists and microbiologists became interested in the role of enzymes in cellular physiology.[8] At this time Woods, an American plant pathologist and physiologist, undertook the study of the relationship between certain enzymes and plant diseases in which a destruction of chlorophyll occurred. He believed that an excess or overactivity of oxidizing enzymes in the plant caused a breakdown of chlorophyll. This in turn produced the chlorotic areas present in various plant diseases. Tobacco mosaic disease was of particular interest to Woods because a mottled or mosaic leaf was the prime symptom of the disease. Furthermore, he knew through familiarity with the work of Mayer, Koning and Beijerinck that no causal micro-organism had been conclusively identified. Unlike previous investigators of the disease, he began his research with the premise that it was not a microbial infection.

Woods reported in 1899 that the amount of oxidase and peroxidase was higher in colorless regions of leaves infected with tobacco mosaic disease. He found that he could produce symptoms of the disease by injecting sterile solutions of peroxidase into young tobacco shoots, but that plants injected with filtered sap from infected plants remained perfectly healthy. (This finding is unaccountable; the filtered sap should have been infectious.) These observations made Woods doubt the infectious nature of the mosaic disease, although he admitted that his "infection experiments have been so few that they cannot have much weight when compared to the positive statements of other investigators, especially Mayer and Beijerinck." (p. 753) Comparing the behavior of oxidizing enzymes to that of Beijerinck's contagium vivum fluidum, Woods discovered that peroxidase also diffused in agar, and that both oxidase and peroxidase remained intact in the soil for several months. Hence he suggested that the agent of tobacco mosaic disease "might be an enzyme, belonging to the oxidases or peroxidases." (p. 752)

Two years later Woods conceded that existing evidence was strongly in favor of the infectious nature of tobacco mosaic disease "under certain conditions," but added that the matter was not settled. In 1902 Woods was still maintaining that "parasites" did not cause the disease. He continued to uphold his thesis that oxidizing enzymes combined with a nitrogen-deficient diet were somehow implicated in its pathogenesis.

Woods' enzyme theory, restricted as it was to plant diseases in which a destruction of chlorophyll occurs, was never intended to be an explanation of plant infections in general, nor of course did it apply to diseases of humans and animals. Even in the few plant diseases to which Woods related his theory, it failed to explain how a nonreproducing chemical substance could continue to cause disease when transmitted through a long series of individuals.

Kurt Gustav Emil Heintzel. In 1900 Heintzel, a recent graduate in natural sciences of the Friedrich Alexander University in Erlangen, Germany, published his doctoral dissertation on tobacco mosaic disease. His research was largely a repetition of previous experiments by Mayer, Ivanovski, Koning and Beijerinck, and contributed very little new information about the disease. Like them he found that sap from tobacco plants with the mosaic disease remained infectious after passage through a porcelain filter and that causal micro-organisms could not be detected in, nor cultured from, infected materials. On the basis of these findings and the dictum of certain botanists that bacteria were incapable of infecting intact living plants (See Chapter 2), he rejected the germ theory as an explanation of the disease. Instead he decided that it arose within the plants themselves, "provoked by specific conditions." (p. 34)

Employing a chemical technique for the extraction of plant enzymes, Heintzel isolated a material from infected tobacco leaves which he claimed produced the mosaic disease when injected into healthy plants. He described it as a soluble, heat-sensitive substance which was unable to multiply outside the plant and which only affected tissues in the process of cell division. It was precipitated by alcohol and was able to destroy chlorophyll with the release of oxygen. "These properties," he concluded, "relate it so closely to the enzymes that one can claim it as an enzyme without any doubt." (p. 44) Having previously decided that the intercellular spaces of leaves infected with tobacco mosaic disease were filled with oxygen (which he evidently considered to be a product of enzymatic activity), he concluded that an oxydase enzyme was responsible for the disease. He apparently reached this conclusion independently of Woods for there is no reference in the dissertation to the latter's work.

In an effort to reconcile the infectiousness of the disease with his enzyme theory of its causation, Heintzel cited a passage from a recent book by Schleichert on another plant enzyme in which the author stated that "uncommonly small amounts of enzyme" might exert a very great effect. Heintzel decided that:

> The ability to transmit the mosaic disease to healthy plants—hence to act infectiously—is explained by the fact that even the smallest amounts of enzyme are able to produce the same effects again and again. (p. 44)

In other words, he suggested, in contrast to Loeffler and Frosch's exclusion of an enzymatic etiology of foot and mouth disease, that an inanimate enzyme might continue to cause disease, despite the dilution it incurred in repeated transfer from plant to plant, if the enzyme retained its pathogenicity at very great dilutions.

The Concepts of the Virus at the Turn of the Century

Both the microbial concept of the virus, describing it as a minute living organism, and the nonmicrobial, as exemplified by Beijerinck's contagium vivum fluidum or the enzymatic theories of Woods and Heintzel, were to have a formative role in the conceptual and experimental approach to viruses in the twentieth century. Viruses were characterized by the properties of filterability and submicroscopic size which revealed very little about their essential nature. Until techniques for the direct determination of viral characteristics were developed, neither concept could be proved nor disproved, nor could viruses be defined in a manner which differentiated them from other microscopically invisible or filterable agents of disease.

It was the microbial concept which was dominant at the close of the century and for several decades thereafter. Research on viral diseases was conducted at this time primarily by medical pathologists and bacteriologists who by and large were advocates of the germ theory. Hence it was reasonable for them to adopt the microbial concept for a disease agent with the properties of infectivity and reproduction. The fact that viruses were filterable and microscopically invisible was explained by assuming organisms to exist which were extremely minute, but which otherwise possessed the properties of ordinary pathogenic micro-organisms. It was possible to attribute the failure to culture viruses *in vitro* to the use of inappropriate nutritive media since infectious agents other than viruses had also failed so far to grow under these conditions.

Aside from having no compelling reason to abandon the germ theory in reference to filterable pathogens, most medical and biological scientists saw distinct disadvantages in the nonmicrobial concept of the virus. As previously mentioned, Beijerinck's idea of an infectious, self-reproducing, noncellular substance was difficult to reconcile with the cell doctrine as well as with the germ theory. Some investigators also realized that to describe an infectious agent as an enzyme, as Woods and Heintzel had done, did not account for its apparently limitless transmissibility. Thus the nonmicrobial concept was adopted by a comparatively small number of scientists who, as in the case of Beijerinck, Woods and Heintzel, were prepared to think in chemical as well as in microbiological terms.

Notes

1. Beijerinck, recognizing that soluble substances could be adsorbed by bacterial filters, later refused to accept Loeffler's experiment with the Kitasato filter as evidence that the agent of foot and mouth disease was cellular. (1900, footnote, p. 299)
2. See Loeffler, 1903, p. 686.
3. The etiological agent of pleuropneumonia is now known to be a mycoplasma, a minute, sometimes filterable, pleomorphic microorganism which grows on artificial substrates. For many years pleuropneumonia was thought to be a viral disease. Because ideas about the nature of its causal agent affected the development of the concept of the virus, a discussion of early research on the disease is relevant.
4. Not until 1903 did Negri describe intracytoplasmic inclusion bodies in the cells of rabid animals. These 'Negri bodies' are now known to contain the rabies virus. *Treponema pallidum*, the spirochete causing syphilis, was discovered by F.R. Schaudinn, in 1905.
5. Beijerinck's first paper on the contagium vivum fluidum was published after Sanarelli's.
6. This is probably the first use of the centrifuge in viral research. However, bacterial filters continued to be the primary tool for separating viruses from cells and larger particles until after the development of the ultracentrifuge by Svedburg in 1923.
7. Studies with the electron microscope show that the tobacco mosaic virus is rod-shaped, approximately 0.300 × 0.015 micron. The virus of foot and mouth disease, one of the smaller viruses, is about 0.024 micron in diameter.
8. For a discussion of biochemical research in the late 19th century, see Kohler.

6

The Development of Concepts of the Virus in the Twentieth Century

This chapter is not meant to be an exhaustive history of viral research in the twentieth century; rather it traces the development of concepts of the virus.

For ease of discussion, the century has been divided into three periods: the early decades, the 1930s and '40s, and the 1950s to the present. The early decades were mainly concerned with research on viral diseases rather than on viruses as such, although viral identification and classification were attempted. The '30s and '40s saw an increasing emphasis on the biochemical analysis of viruses and the introduction of techniques new to viral research, such as electron microscopy and x-ray crystallography. These were then applied, particularly in the second half of the century, to elucidate the essential nature of viruses at the morphological and molecular levels.

It would be misleading to isolate a single event as the point of origin of the present concept of the virus. Reference to Table II indicates that its evolution has been a cumulative and continuous one.

Table II

LANDMARKS IN VIRAL RESEARCH

Date	Scientist	Accomplishment*
1798	E. Jenner	Empirical development of smallpox vaccine.
1841	W. Henderson R. Paterson	Independent microscopical observations and descriptions of the inclusion bodies of molluscum contagiosum.

*The accomplishments are described in modern terms and in the light of present virological knowledge; some of the original investigators did not use these terms nor did they always recognize the full significance of their work.

1886	J. Buist	Microscopical observation and description of the elementary bodies of vaccinia and variola.
1892	D.I. Ivanovski	First description of the filterability of a virus (tobacco mosaic).
1898 (Mar.)	F. Loeffler and P. Frosch	First description of the filterability of an animal virus (foot and mouth).
1898 (Apr.)	E. Nocard, E. Roux *et al.*	Discovery of the mycoplasma (of bovine pleuropneumonia).
1898 (June)	G. Sanarelli	Discovery of the virus of myxomatosis.
1898 (Nov.)	M.W. Beijerinck	Description of the tobacco mosaic virus as a *contagium vivum fluidum*; discovery of its exclusively intracellular mode of reproduction.
1900	J. McFadyean	Discovery of the virus of African horse sickness.
1901	W. Reed *et al.*	Discovery of the virus of yellow fever.
1901	A. Lode and J. Gruber	Discovery of the virus of fowl plague.
1902	A. Borrel	Discovery of the virus of sheeppox.
1902	M. Nicolle and Adil-Bey	Discovery of the virus of cattle plague.
1902	E. Marx and A. Sticker	Discovery of the virus of fowlpox.
1903	D.I. Ivanovski	Microscopical observation and description of crystalline inclusions of tobacco mosaic virus in diseased tobacco leaves.
1903	A. Negri	Description of the inclusion bodies of rabies.
1903	P. Remlinger and Riffat-Bey	Discovery of the virus of rabies.
1905	M. Juliusberg	Discovery of the virus of molluscum contagiosum.
1907	P.M. Ashburn and C.F. Craig	Discovery of the virus of dengue fever.
1909	S. Flexner and P.A. Lewis	Discovery of the virus of poliomyelitis.
1911	J. Goldberger and J.F. Anderson	Discovery of the virus of measles.

1911	P. Rous	Transmission of a tumor (of chicken sarcoma) by means of a cell-free filtrate.
1913	E. Steinhardt, C. Israeli and R.A. Lambert	One of the first propagations of a virus in tissue culture.
1915	F.W. Twort	Discovery of a virus infecting and lysing bacteria.
1917	F. d'Herelle	Rediscovery of Twort's virus which d'Herelle named 'bacteriophage.'
1921	J. Bordet and M. Ciuca	One of the first descriptions of lysogeny.
1930	M. Theiler	Demonstration of the usefulness of mice in viral research.
1931	W.J. Elford	Description of the use of graded collodion membranes for the determination of viral size.
1931	A.M. Woodruff and E.W. Goodpasture	Cultivation of a virus (fowlpox) in developing chick embryos.
1934	C.D. Johnson and E.W. Goodpasture	Discovery of the virus of mumps.
1935	W.M. Stanley	The first artificial crystallization of a virus (tobacco mosaic).
1935	M. Hoskins	The first experimental demonstration of interference between animal viruses.
1936	M. Schlesinger	Demonstration that bacteriophage consist of nucleoprotein.
1938	Y. Hiro and S. Tasaka	Discovery of the virus of rubella.
1939	G.A. Kausche, E. Pfankuch and H. Ruska	First use of the electron microscope for the visualization of a virus.
1941	G.K. Hirst	Description of the hemagglutination-inhibition test.
1946	M. Delbrück and W.T. Bailey	Discovery of genetic recombination in bacteriophage.
1948	G. Dalldorf and G.M. Sickles	Discovery of the first Coxsackie virus.*

*See Table III for some of the other viruses discovered by isolation in tissue culture.

1949	J.H. Enders, T.H. Weller and F.C. Robbins	*In vitro* cultivation of poliovirus in non-neural tissues.
1952	R. Dulbecco	Description of the plaque method for the quantitative study of viruses.
1952	N.D. Zinder and J. Lederberg	Description of transduction, the transference of genetic characters between bacteria by means of bacteriophage vectors.
1952	A.D. Hershey and M. Chase	Demonstration that DNA is the infective component of bacteriophage.
1953	J.D. Watson and F.H.C. Crick	Description of the molecular structure and replicative mechanism of DNA.
1955	F.L. Schaffer and C.E. Schwerdt	The first artificial crystallization of an animal virus (poliovirus).
1955	H. Fraenkel-Conrat and R.C. Williams	Reconstitution of active tobacco mosaic virus from its inactive protein and nucleic acid components.
1956	A. Gierer and G. Schramm H. Fraenkel-Conrat	Demonstration that RNA is the infective component of tobacco mosaic virus.
1957	A. Isaacs and J. Lindenmann	Discovery of interferon.
1962	T.O. Diener	Discovery of the viroid.
1970	D. Baltimore	Discovery of an RNA polymerase in an RNA virion (vesicular stomatitis virus).
1970	D. Baltimore H.M. Temin and S. Mizutaki	Independent discovery of an RNA-dependent DNA polymerase in RNA tumor virions (Rauscher mouse leukemia and Rous sarcoma virus).

The Early Decades. It is not surprising that research in the early decades focused primarily on viral diseases rather than on the precise nature of viruses which could not be visualized nor grown *in vitro*. The manifestations of viral diseases, some of the major infectious diseases of the world, were far easier to identify than were the causal agents. Furthermore, there was at this time no unified approach to viral research. Viral investigators were divided into animal and plant pathologists and bacteriologists; later came the bacteriophage investigators. All worked relatively independently and with somewhat different conceptual approaches.

Investigation of human and animal diseases was conducted largely by medical pathologists and bacteriologists who thought in terms of the cell and germ theories. Consequently they were predisposed to assume that these minute, filterable agents were cellular (microbial) in nature. As recently as the last decade of the nineteenth century, plant pathologists had recognized the bacterial etiology of some plant diseases. (See Chapter 2) They were more used to thinking in terms of plant enzymes and chemical mediators and hence were more inclined towards a nonmicrobial concept of the virus.

The investigators of bacteriophage (viruses that infect bacteria) were mainly interested in the therapeutic use of phage as antibacterial agents. They recognized the futility of using current techniques for direct investigation of these viruses for which some held microbial and others nonmicrobial concepts.

Although radical changes in concepts of the virus did not occur in these years, it would be a mistake to assume that there were no advances in viral knowledge. For example, yellow fever (1901), poliomyelitis (1909), and many other diseases were found to be caused by filterable and invisible agents whose specific nature and mode of replication raised much interest and many questions. Several hypotheses were proposed to explain the nature of these puzzling infectious agents.

Nonmicrobial Concepts of the Virus. Nonmicrobial concepts continued to play a role, albeit a subordinant one, in the early years of the twentieth century.

In 1901 A. Lode and J. Gruber of the Institute of Hygiene at the University of Innsbruck offered two hypotheses for the nature of the agent of fowl plague which they had found to be filterable.[1] The easier to accept, they said, described the "mysterious germs" as consisting of "half-fluid protoplasm whose various changes in form we have been able to follow in the microscope." (p. 603) Presumably it was by this ameboid movement that the agent passed through the filter. Their second hypothesis was that

> the virus [infectious agent] of this puzzling infection is not at all corporeal, but rather a soluble, reproducing substance of somewhat enzyme-like character which acts by means of the processes of decomposition which it causes in the animal body, without itself being destroyed. (p. 604)

They admitted that such a substance was hard to imagine. Joest agreed: "An infectious agent in the form of a soluble, enzyme-like, reproducing substance is . . . absurd." (p. 420)

The following year, on the basis of Loeffler and Frosch's work on foot and mouth disease and his own filtration experiments, Lode changed his opinion about the nature of the fowl plague virus. He decided that there was no need to postulate the involvement of an unusual infectious agent and abandoned the nonmicrobial for the microbial concept.

Eugenio Centanni, a pathologist at the University of Ferrara, was led by his research on the same outbreak of fowl plague to raise the possibility of a nonmicrobial infectious agent. Like Lode and Gruber, he found that it was both filterable and microscopically invisible. His remarks demonstrate an awareness of the problem of determining the fundamental nature of the virus:

> Because of the unsatisfactory state of current research methods, we are of course going to lay aside the question of whether in these viruses the reproduction of living organisms or complicated chemical molecules is involved, or even elements which stand on the border between one realm and another. (p. 198)

Thus realistically taking into account the inadequacy of current research methods for clarifying the nature of the virus, he turned to more accessible problems: the discussion of those infectious diseases known by 1901 to be caused by filterable disease agents (foot and mouth disease, tobacco mosaic disease, African horse sickness, bovine pleuropneumonia and fowl plague).

In 1905 F.W.T. Hunger of the Botanical Institute at the University of Utrecht took a definite stand in favor of the nonmicrobial concept of the virus. Like Woods, he believed tobacco mosaic disease to be a physiological disturbance due to unfavorable growth conditions. He ruled out bacteria as causal agents on the basis of his finding that the structures which Ivanovski had designated as the bacteria (zooglea) of tobacco mosaic disease could be chemically dissolved without affecting the cells of the tobacco plant. Hunger reasoned that these structures were not bacteria because they, being cells, would not have dissolved. He also decided that Beijerinck's idea of a contagium vivum fluidum was inadequately substantiated and he rejected the Woods-Heintzel oxidizing enzyme theory because it could not be reconciled with the infinite transmissability of the agent of tobacco mosaic disease. (1905a, pp. 415–416) Instead he proposed that the pathogen was a nonliving "phytotoxin." This toxin, normally a harmless metabolic product of the plant cell, caused physiological disturbances, such as tobacco mosaic disease, when it accumulated as a result of "greatly elevated plant metabolism." The toxin could then enter normal cells where it induced production of more toxin by a "physiological contact action." Thus Hunger believed that the transmissibility of the causal agent could be explained by the fact that "the toxin of the mosaic disease has the property of acting in a physiologic-autocatalytic fashion." (1905a, p. 417)

Hunger's theory of the cause of tobacco mosaic disease, which was as speculative as that of his predecessors, is of interest because, as Wilkinson has pointed out (1976), he clearly expressed his belief in the pathogen's autonomous origin within the plant. (1905a, p. 416) This idea was later applied to other viruses, particularly the tumor-producing and bacterial viruses. (See (p. 84 and p. 85 below) It was finally eliminated in the 1950s when a more precise mechanism of viral replication became known.

The work of Mrowka, a veterinarian in the German military, is yet another example of an early nonmicrobial approach to the nature of the virus. In a paper of 1912 entitled "The virus of fowl plague a globulin," he described the results of experiments on the virus using chemical and physical methods from the currently active field of protein chemistry. Finding that the infectious agent was not precipitated after "hour-long" centrifugation, he concluded, "absurd as the thought might first appear, the virus must be dissolved in the fluid in some form." (p. 251) Further evidence for the solubility of the virus, Mrowka believed, was the fact that it was able to penetrate capillary walls. In precipitation experiments on infectious serum, he found that the virus was always restricted to the insoluble globulin fraction. He concluded that it was a globulin (p. 266) and hence "had every reason to doubt the vital properties of the virus." "Knowledge of the nature of the virus," he continued, "definitely supports the view that the filterable virus is limited to the animal[2] body and very probably to a certain type of protein in the living body and can multiply only in the living organism." (p. 267) Mrowka's stated intention to publish further on this subject was never fulfilled.

In 1914 P. Andriewsky, a Russian veterinarian working at the Pasteur Institute in Brussels, suggested that the agent of fowl plague was a noncellular substance akin to Beijerinck's contagium vivum fluidum. He arrived at this conclusion as a result of attempts to calculate the dimensions of the agent after passing it through the graded ultrafilters invented by Bechhold. The results indicated that the infectious agent of fowl plague consisted of "micelles" or "molecules"[3] smaller than the hemoglobin molecule which was currently thought to have a diameter of 2.3–2.5 micromicrons.[4] Consequently Andriewsky concluded that "the virus cannot be formed of cells similar to the animal and plant cells known at present."

The Microbial Concept of the Virus. The work on African horse sickness by John McFadyean of the Royal Veterinary College, London, is typical of the many applications of the microbial concept in early viral research. In 1900 he reported that the causal agent was easily filtered and referred to it as the "microbe" or "bacterium of horse-sickness." He apparently assumed it to be a minute micro-organism, similar to those described by Loeffler and Frosch and by Nocard and Roux, whose research he cites. (p. 20)

In a review of the "ultravisible viruses" published eight years later, McFadyean's adherence to the microbial concept is implicit:

> . . . it is necessary to point out that there is no wide gap between the visible and the ultravisible microbes. It may be said to have been *a priori* probable almost to the point of certainty that in nature there must be bacteria much smaller than the fowl cholera bacillus or the bacillus of influenza which are the smallest of the microbes that have been made distinctly visible with the microscope; further, that there could be no wide gap between the visible and the ultravisible bacteria. (p. 62)

Nevertheless, McFadyean was puzzled by the failure of the "ultravisible viruses" to grow in artificial media:

> One remarkable feature common to the whole of the ultravisible viruses is that they have hitherto resisted all attempts to cultivate them in artificial media outside the body. Assuming that these viruses are bacterial in their nature, one has great difficulty in understanding why this should be so, for one does not see why the ability of a bacterium to grow under artificial conditions should in any degree be dependent upon its size. (p. 240)

Viral Inclusion and Elementary Bodies. A cellular inclusion is a microscopically visible virus associated with material produced by the host cell in reaction to its presence. Cellular inclusions had long been observed in association with certain infectious diseases (See Chapter II) and they became the focus of considerable debate towards the end of the first decade and for some time thereafter. Visible evidence of the presence of viruses they were. But were they the actual pathogens; were they protozoa in an intracellular stage of their life cycle; or were they simply cellular reaction material?

Stanislaus von Prowazek, a Bohemian microscopist at the Institute of Tropical Hygiene in Hamburg, suggested in his theory of the 'Chlamydozoa'[5] that the inclusions were of dual composition. He believed that they were filterable micro-organisms which developed intracellularly and which were enveloped in a mantle of cellular reaction material. He was unsure about their classification, but thought that they were more closely related to the protozoa than to the bacteria.

In 1909 the Austrian dermatologist Benjamin Lipschütz, concerned with finding unifying characteristics for at least some of the viruses, placed the microscopically visible and filterable ones in a separate group to which he gave the name 'Strongyloplasma.'[6] He described them as round, darkly-staining bodies which were usually obligate parasites of the host cell. These bodies, as is now known, were microscopically visible viruses.

By 1913 Lipschütz had listed forty-one diseases thought to be caused by filterable infectious agents, of which sixteen were microscopically visible as inclusion or elementary bodies. The presence of these bodies, and the varied form they took, provoked questions about the identity of the infectious agents, especially their relationship, if any, to the bacteria and protozoa. These questions were eventually resolved in 1929, when C.E. Woodruff and E.W. Goodpasture showed that the inclusion bodies of fowlpox contain elementary bodies which are actually the virus.

Lipschütz's Strongyloplasma differed from von Prowazek's Chlamydozoa only in the fact that the former were not invariably associated with cellular reaction material. Their observations were convincing demonstrations of the fact that filterability and submicroscopic size are not synonymous.

As far as the concept of the virus is concerned, the research on inclusion bodies during the first decades of this century made few contributions to knowledge about the intrinsic properties of viruses. However, it did demonstrate the possibility of grouping viruses on the basis of common biological properties (such as their manifestation as intracellular inclusions or elementary bodies), rather than on purely physical ones, such as filterability and submicroscopic size. In subsequent attempts to systematize the viruses, other biological criteria, such as the nature of the host organism or their exclusive affinity for living cells, were used.[7]

Viral Tumors. In 1908 V. Ellermann and O. Bang, working in Copenhagen, discovered that leukemia of hens could be transmitted by cell-free filtrates. They did not describe the nature of the causal agent except to say that it was probably "organized," i.e. cellular. Three years later Peyton Rous at the Rockefeller Institute of Medical Research in New York demonstrated conclusively that a sarcoma of fowls could be transmitted in the same manner. The fact that the pathogen was filterable suggested to many that it might be a virus.

These findings stimulated interest in the nature of tumor viruses and the possible relationship between them and cancer—an issue which is still current. Were they alive or did they arise *de novo* in the tumor? Did the virus cause the tumor, or the tumor the virus?

Rous maintained that the sarcoma virus was animate. Others believed it to be an inanimate substance derived from the injured host cell and somehow able to regenerate itself through its successive action upon normal cells. W.E. Gye ascribed the cause of the Rous tumor to two factors, a labile chemical substance produced by tumor cells and a virus. He suggested that the chemical substance in an unknown way rendered the host cells susceptible to infection by the virus which alone was unable to initiate the malignant transformation. (p. 116)

Bacteriophage. In 1915 the British bacteriologist, Frederick William Twort, superintendent of the Brown Institution at the University of London, published a short paper entitled "An investigation on the nature of the ultramicroscopic viruses." This paper marks the beginning of a new phase of viral research which was to have significant effects on the concept of the virus. Twort described a filterable principle which caused lysis of bacterial colonies, a condition which could be transmitted to fresh cultures for an indefinite number of generations. This was the first indication that bacteria, as well as plants and animals, are susceptible to disease. Uncertain about the nature of the lytic substance, he suggested that it might be produced by the bacteria but also maintained that "the possibility of its being an ultramicroscopic virus has not been *definitely* disproved. . . ." Perhaps because of Twort's uncertainty and the interruption of his work by the outbreak of World War I, his contribution passed virtually unnoticed for a time.

Two years later Felix d'Herelle, a French Canadian working on dysentery at the Pasteur Institute in Paris, rediscovered Twort's lytic principle. He observed that filtrates from lysed cultures of the dysentery bacillus caused further lysis when transferred to fresh cultures. Because this phenomenon could be continued indefinitely and because the lytic principle actually increased in the course of passage, d'Herelle decided that the lysis was caused by "an invisible microbe endowed with antagonistic properties against the Shiga dysentery bacillus." He also noted that, because the "microbe" did not grow in any medium in the absence of the dysentery bacillus, it was an obligate parasite of various species of the bacillus. To this unique agent he gave the name 'bacteriophage' (literally bacteria-eater), classifying it with the 'filterable viruses.'

Like the Rous sarcoma virus, d'Herelle's work on bacteriophage aroused wide interest and intense debate. Not everyone agreed with d'Herelle that bacteriophage were living organisms. Some held that they were bacterial enzymes which stimulated bacteria to synthesize more enzyme, eventually resulting in the destruction of the bacteria. Others insisted that bacteria spontaneously generated bacteriophage.

The attraction of bacteriophage for scientists of this period was their possible use as therapeutic agents for treating bacterial infections. Although this proved not to be feasible, bacteriophage assumed an increasingly prominent place in viral research. They were employed to study the relationship of the virus to the host cell and later in work on the nature of the gene. By 1930 E.W. Schultz of Stanford University could say that "much of the work carried out in recent years on viruses in general can be credited in a large measure to the stimulus which has grown out of the studies on the nature of bacteriophage." (p. 424)

The Problem of Characterizing the Virus. As knowledge about filterable infectious agents accrued in the early years of this century, it became apparent that they varied considerably in size, resistance to chemical and physical agents, and pathogenic effect.

Indeed it was suggested, as we have seen, that they might not be a homogeneous biological group and that some should be classified with the protozoa or with the bacteria. A few investigators considered existing categories to be inadequate for classifying filterable infectious agents. For example, S.B. Wolbach, an American pathologist, remarked in 1912:

> It is quite possible that when our knowledge of filterable viruses is more complete, our conception of living matter will change considerably, and that we shall cease to attempt to classify the filterable viruses as animal or plant.

We now know that some rickettsiae[8] and pleuropneumonia-like organisms were included amongst what was currently termed the 'filterable viruses.' Although these micro-organisms may have contributed to the confusion, they were not the basic problem; consideration of viral size alone was not an adequate basis for biological classification.

The disadvantages of filterability and submicroscopic size as criteria for characterizing viruses were recognized very early in the twentieth century. First, filterable agents were not necessarily submicroscopic, as Nocard and Roux had discovered earlier in their research on bovine pleuropneumonia. With this work in mind, Eugenio Centanni proposed in 1902 that 'filterable' rather than 'ultravisible' was a more useful description of viruses. (p. 198) A year later Emile Roux made the same point: "Just because a microbe traverses a filter does not mean that it deserves for that reason to be called invisible. . . ." (p. 55)

Second, it was found that submicroscopic disease agents did not invariably pass through bacterial filters. In these years it became well recognized that the procedure of filtration involved more than just the relationship between the pore diameter of the filter and the size of the particles to be filtered. Other factors to be considered in interpreting the results of the filtration of small particles were frequently discussed. For example, in 1911 the Hungarian pathologist Robert Doerr read a paper to a meeting of microbiologists in Dresden on the problems of filtering viruses.[9] The nature of the medium, the forces of molecular attraction and capillarity, and the duration and pressure of filtration were some of the factors he discussed.

The discovery in 1898 by Nocard and Roux of the filterable pleuropneumonia organism, which was just visible at the lower limit of microscopic visibility, established that there need be no sharp break between microscopic and submicroscopic entities, as formerly imagined. Instead it was now believed that there was a continuous gradation in size from microscopic to submicroscopic organisms and particles. It was recognized that filterability and ultrascopic size were technique-determined physical characteristics which provided little information about intrinsic biological properties. It followed that neither filterability nor microscopic invisibility were adequate criteria by which to differentiate viruses from all other types of infectious agents. Nonetheless, 'filterable virus' became the preferred term in the early decades of the twentieth century.[10]

Attempts to culture viruses on artificial substrates continued to be unsuccessful. Because most investigators assumed that they were a type of micro-organism, they expected them to grow *in vitro* once the right conditions had been met. The inability of viruses to replicate outside a living cell, which would eventually be designated as a distinguishing property, was not widely recognized at this time.

However, John McFadyean and a few others suggested that viruses might be obligate parasites restricted in their reproduction to the body of a living host. (1908, p. 241) And André Philibert of the Faculty of Medicine in Paris wrote in 1924 that "this exclusive affinity for the living cell is the fundamental characteristic of viruses . . . ," one which distinguishes them from all microbes.[11]

The solution to the problem of viral classification was to find a biological property which would clearly identify the viruses as a group. This was difficult, if not impossible, to achieve at that time. The microbiological procedures of filtration, microscopy and *in vitro* cultivation produced only negative evidence; viruses were not retained by bacterial filters, were usually not visible in the microscope and would not grow on artificial media.

The 1930s and 1940s

The two areas of primary interest in viral research in the 1930s and '40s were the development of new techniques and the determination of the nature of the virus. The latter was implicit in all aspects of viral research. It was most immediately confronted in the work on viral size and biochemical composition.

The Determination of the Nature of the Virus

Viral size. The results of research on viral size revitalized the continuing argument over the animate versus the inanimate nature of viruses. Scientists debated the minimal degree of organization which could reasonably be attributed to a living organism. How many protein molecules were needed to form the basic living unit? Could a virus the size of a large molecule have all the characteristics of life? What precisely was the difference between living and nonliving?

In 1931 I.A. Galloway and W.J. Elford of the National Institute for Medical Research in London determined the size of the virus of foot and mouth disease by filtering it through collodion membranes with pores of known diameter which Elford had devised specifically for the purpose of viral measurement (Elford, 1931). Their finding that it was only 0.008 to 0.012 micron in diameter was interpreted by some observers as an indication that viruses were nonliving entities; they doubted that particles approximating the size of some protein molecules could be complex enough to be truly alive. For example, Sir Henry Dale, the future Nobel laureate in physiology, commented in 1931 that

> the dimensions assigned to the units of some viruses, representing them as equal in size to mere fractions of a protein molecule, might well make one hesitate to credit them with the powers of active self-multiplication. (p. 601)

It was later learned that their figure was too small.[12]

Elford and Andrewes discovered that the poxviruses were many times the size of the virus of foot and mouth disease; there was also a wide range of size within the phage. It was becoming more apparent that viruses were a heterogeneous group. This realization made it possible for some investigators of this period to suggest that the larger viruses were minute micro-organisms and the smaller ones infectious chemical substances. Even Thomas Milton Rivers, a pediatrician turned pathologist at the Rockefeller Institute in New York City and previously a staunch advocate of the microbial concept, admitted in 1932 that some viruses "may be minute organisms, while others may represent forms of life unfamiliar to us, while still others may be inanimate transmissible incitants of disease."

If this were true, then the viruses could not be regarded as a distinct biological group but merely a collection of diverse pathogenic agents accidentally sharing the physical properties of filterability and minute size. Rivers came to a decision befitting his training in medical pathology:

> The confused state of our knowledge of the viruses at the present time makes it exceedingly difficult to determine the nature of these active agents. The easiest way out of the dilemma, however, would be the acceptance of the presumptive evidence that viruses are minute organisms. (1932)

Most of his medical colleagues would have agreed with him.

Biochemical Composition

Viral research for the first third of the twentieth century had been directed towards the investigation of viral diseases and hence was the concern primarily of pathologists. However, in 1935 Wendell Meredith Stanley, a biochemist at the Rockefeller Institute for Medical Research in Princeton, New Jersey, published a paper which was to change its course and provided support for the nonmicrobial concept of the virus.

He announced that he had crystallized the virus of tobacco mosaic disease.[13] This he achieved by using techniques devised for crystallizing proteins. Finding that the "crystalline protein" was considerably more infectious than suspensions of either diseased tobacco leaves or the twice-frozen juice of infected plants, and that a 1 to 1,000,000,000 dilution of the crystals usually produced typical tobacco mosaic disease, he concluded:

> Tobacco-mosaic virus is regarded as an autocatalytic protein which, for the present, may be assumed to require the presence of living cells for multiplication.

In 1946 Stanley received the Nobel Prize in Chemistry for his crystallization work.

The ability to obtain plant viruses in a highly purified form initiated an era of intense biochemical research. By linking biological activity with a tangible chemical entity, Stanley demonstrated that viruses could be studied from a biochemical as well as from the traditional microbiological standpoint.

Investigators previously interested in viruses primarily as agents of infectious disease now began to explore the composition of the virus and the phenomena of viral infection and replication. The new biochemical approach to viruses revived the old question of whether they were living or nonliving. In general, as might be expected, pathologists and microbiologists held that they were living organisms while physical scientists tended to argue that they were inanimate chemical substances.

Stanley's observation that tobacco mosaic virus was a large protein molecule was soon modified by the British scientists, F.C. Bawden and N.W. Pirie of the Biochemical Laboratory at Cambridge University, who announced in 1937 that the virus was in fact a nucleoprotein.

In the same year Max Schlesinger, working at the National Institute for Medical Research, London, reported that purified preparations of bacteriophage consisted mainly of nucleic acid and protein. The nucleic acids were thought to play a relatively minor role in viral metabolism. Their genetic significance was not widely recognized for many years because it had been thought, since the early years of the twentieth century, that proteins were the likeliest carriers of genetic information. Their twenty or more amino acids seemed more suited than the four nucleotides of nucleic acids to form the code of the hereditary message. Not until 1952 was it commonly accepted that nucleic acids are the basic genetic substance. (See pp. 100–102 below)

Bacteriophage were particularly suited for the study of viral composition, infection and replication. As Emory L. Ellis, an associate of Delbrück at the California Institute of Technology, pointed out retrospectively in 1966, they had four distinct advantages, as compared with animal and plant viruses, for studying viral behavior:

1. The quantitative assay of phage grown in bacteria was more rapid than most biological assays using animal or plant hosts.
2. The individual phage or the average behavior of a larger population could be studied.
3. The investigator had nearly absolute control over the experimental environment.
4. Phage provided excellent material for elucidating the process of genetic replication.

Because of these advantages, which were enhanced by advances in electron microscopy, radioactive tracing and chemical analysis, scientists trained in a variety of disciplines, including genetics, physics, biophysics, biochemistry and cytology, became interested in bacteriophage as a means for studying basic biological processes.

An extremely important result of this research was the discovery that viruses, unlike cellular organisms, do not reproduce by binary fission. Instead they have a unique form of replication within the host cell.

A number of prominent physicists, among them the Austrian Erwin Schrödinger and the German Max Delbrück, were attracted to biology, particularly to viral research. Donald Fleming attributes this attraction to two basic mòtives. First, these physicists wanted to leave war-torn Europe and the oppressive ethical problems involved in work on the atom bomb. Second, they wanted to participate in "life-enhancing" research. They believed that the "ultimate secret of life," namely the nature and mode of replication of the gene, was within immediate scientific reach and could be explained by physical and chemical laws.

One of the men in the vanguard of the physicists-turned-biologists was the previously mentioned Max Delbrück. He left Germany for the California Institute of Technology in 1937 and took up research on bacteriophage with the objective of discovering the physical basis of heredity. In 1939 Ellis and Delbrück contributed an invaluable technique to phage research, the one-step growth experiment.[14] It was now possible to do highly quantitative studies of the phage-bacterium interaction. By 1940 Delbrück was collaborating with the Italian biologist, Salvador Luria, and the American biologist, A.D. Hershey. They were also interested in phage as a key to the genetic mechanism. Together they oriented phage research towards genetic problems and formed the nucleus of the celebrated "phage group" of the 1940s. For their work on bacteriophage these three men won the Nobel Prize for Physiology and Medicine in 1969.

The investigation of the phage, performed by scientists from various disciplines, widened the scope of viral research. It set a new standard for precise, quantitative results and demonstrated the value of a cooperative, multidisciplinary approach. These investigators performed important work on the phenomena of viral infection, mutation and genetic recombination. The technical and theoretical milieu of the phage school provided impetus for the formation of molecular biology[15] and for the eventual elucidation of the structure and function of DNA.

The Origin and Nature of Viruses

The question of the origin of viruses was actively debated in this period. Some maintained that viruses were descendants of microorganisms which had gradually become more dependent on their hosts, losing by "retrograde evolution" certain structures and functions. (Green) Others believed that viruses had originally been components of cellular organelles which had evolved towards greater independence from the cell. (Darlington) Both theories are difficult to prove and the question of the origin of viruses is still unanswered.

With regard to the nature of viruses, biochemical findings appeared to support the idea that viruses are very large molecules, a refinement of the nonmicrobial concept of the virus. Yet it was also true that the ability of viruses to multiply and to infect were properties traditionally associated with the living state. Hence they possessed both animate and inanimate characteristics. Stanley circumvented this dilemma when he stated in 1939 that:

> . . . there is in reality a continuum from simple to complex structures, from molecules to organisms, and after all there is no great difference between the two.

The Application of New Techniques in Viral Research

During these years filterability continued to be used as a criterion for classifying infectious agents as viruses, even though there was considerable dissatisfaction with the standardization and significance of the procedure. Sir Henry Dale remarked in 1931:

> The crude qualitative distinction between the filterable and nonfilterable agents of infection has long since ceased to have any real meaning. There is no natural limit of filterability. A filter can be made to stop or to pass particles of any required size. (p. 600)

Because filterability was not by itself an adequate means for classifying viruses, various types of filter-passing pathogens continued to be identified as 'filterable viruses.' In 1928 Rivers drew up a list of over sixty-five diseases which different investigators considered to belong to the filterable virus group. (See Table III) Included in the list were diseases now known to be caused by rickettsiae or bacteria.

By the mid-1920s a number of investigators had cited the exclusive dependence of viruses on the cell for reproduction as one of their salient features. However, some investigators continued to insist that they would one day be cultivated on artificial substrates; a few even claimed success in these endeavors. But in the 1930s and 40's the more common view was that viruses required living cells in order to multiply. Evidence for the intracellular site of viral replication came from two active areas of contemporary viral research, namely the work on bacteriophage and inclusion bodies. In 1930, in a statement that presages to some extent the phenomenon of nucleic acid replication, Schultz suggested that viral reproduction

> might be accomplished through the medium of the host cell deranged in some way by an appropriate agent setting up a disturbance of such a nature as to cause the cell itself to elaborate more of the selfsame principle capable of affecting new cells. (p. 430)

Tissue Culture and the Embryonated Egg. The dependence of viruses on living cells for their reproduction was the basis of these two techniques which were to become extremely important in research on viral diseases. In 1913, Edna Steinhardt, C. Israeli and R.A. Lambert had demonstrated that the vaccinia virus could be grown in tissue cultures of rabbit or guinea pig cornea. A similar achievement had been reported the same year by C. Levaditi who used *in vitro* cultures of spinal ganglia to grow poliovirus. Tissue culture circumvented the complexities of research on the intact animal. Yet because of the problems involved in maintaining aseptic conditions, the technique was not widely used in viral research until the 1950s. (See p. 96 below)

In 1931 Alice Miles Woodruff and Ernest W. Goodpasture, pathologists at Vanderbilt University, introduced another method for cultivating viruses. They reported that the virus of fowlpox could be grown on the chorioallantoic membrane of developing chick embryos;[16] lesions containing the virus appeared on the membrane after inoculation. Thus they introduced into viral research an experimental host which, in contrast to laboratory animals, was cheap and readily obtainable. It had a variety of membranes susceptible to infection by different viruses and was relatively easy to keep sterile.

The use of living cells for the culture of viruses necessitated modification of Koch's postulates; his idea that the etiological

Majority of the Diseases Which Have Been Placed in the Filterable Virus Group by Different Workers

Bacteriophagy

Mosaic diseases of plants (infectious chloroses)

Sacbrood
Wilt of European nun moth
Wilt of gipsy-moth caterpillar
Jaundice of silkworms

Epizoötic of guinea pigs
Hog cholera
Cattle plague (Rinderpest)
Pernicious anemia of horses
Salivary-gland disease of guinea pigs
Virus III infection of rabbits
Foot-and-mouth disease
1. Type A
2. Type O
Vesicular stomatitis of horses
Paravaccinia (No report on filtration)

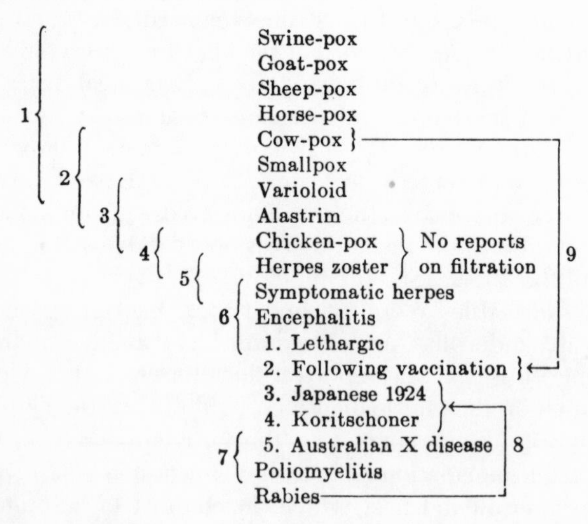

Borna disease
Fowl plague and plague of blackbirds
Guinea-pig paralysis

Distemper of dogs
Infectious papular stomatitis of cattle
Molluscum contagiosum
Warts
Contagious epithelioma (fowl-pox)

1. Chickens
2. Pigeons

Infectious myxomatosis of rabbits
Rous sarcoma of chickens
Leukemia of chickens
Lymphocystic disease of fish
Epithelioma of fish
Carp-pox
} No reports on filtration

Trachoma* and inclusion blenorrhea
Mumps (According to Kermorgant, a spirochetal disease)
Agalactia (According to Bridré, a bacterial disease)

Measles (rubeola)
German measles (rubella) (No report on filtration)

Grippe (influenza) (According to Olitsky and Gates, a bacterial disease)
Common colds

A. Insect-borne	a.	Nairobi disease of sheep Catarrhal fever (blue tongue) of sheep African horse sickness Pappataci fever Dengue fever Yellow fever (According to Noguchi, a spirochetal disease)
	b. Rickettsia diseases	Typhus fever Trench fever Rocky Mountain spotted fever Heartwater disease Flood fever of Japan
B. Bacterial diseases		Oroya fever and verruga peruviana Pleuropneumonia Avian diphtheria Scarlet fever

---

* According to Dr. Noguchi, trachoma of American Indians is a bacterial disease.

Table III From "Some general aspects of filterable viruses" by T.M. Rivers in *Filterable Viruses*, T.M. Rivers (ed.), Baltimore: The Williams and Wilkins Co., 1928.

agent had to be isolated and cultured on artificial media was no longer invariably required to establish the cause of an infectious disease. In 1937 Rivers stated:

> The idea that an infectious agent must be cultivated in a pure state on lifeless media before it can be accepted as the proved cause of a disease has also hindered the investigations of certain maladies, inasmuch as it denies the existence of obligate parasitisms, the most striking phenomenon of some infections, particularly those caused by viruses.

Electron Microscopy. Direct visual evidence of submicroscopic viruses was obtained for the first time in 1939 when G.A. Kausche, E. Pfankuch and H. Ruska published electron micrographs of the virus of tobacco mosaic disease. Resolution was later enhanced by the use of the technique of metallic shadowing, introduced by Robley Williams and Ralph Wyckoff in 1944. A heavy metal such as uranium was vaporized and allowed to settle on a virus preparation. The shadows cast by the plated virus particles enhanced the contrast and hence the resolution of electron micrographs.

Electron microscopy permitted the measurement of viral size, thereby supplementing the indirect methods of ultrafiltration and high-speed centrifugation. (The latter technique was perfected in the 1930s by the Swede Theodor Svedberg and, independently, by Ernest E. Pickels at the Rockefeller Institute.) As the electron microscope and associated techniques were improved, detailed information gathered about the morphology and function of viruses. For example, Thomas F. Anderson's electron micrographs of the '40s and early '50s revealed the distinctive morphology of phage. Furthermore, their empty protein coats seen attached to bacterial cell walls suggested that it was only the nucleic acid component of phage which functioned in their replication within the cell. Anderson's micrographs also demonstrated another phase of the growth cycle—bacterial lysis with the liberation of phage particles.

The Hemagglutination-Inhibition Test. The agglutination of red blood cells by viruses, first reported in 1941 by George Keble Hirst, provided the basis for the hemagglutination-inhibition test, a means for differentiating viral strains and for diagnosing specific viral infections (at least for those viruses which agglutinate red blood cells). This agglutination reaction was also used to study virus adsorption at the cell surface, the earliest stage of viral infection.

X-ray Crystallography. By the 1940s x-ray crystallography was being used in viral research. This technique and those of electron microscopy and immunological assay emphasized the macromolecular aspects of viruses and made possible the accurate determination of viral structure and composition. X-ray crystallography was to play a major role in unravelling the structure of the DNA molecule in the early '50s.

1950 to the Present: The Formulation of a Precise Concept of the Virus

Increased activity in virological research after midcentury resulted in the rapid development of experimental techniques adapted specifically to the investigation of viruses. The knowledge gained through their application was used to develop a more precise concept of the virus. This was very important in the unification and maturation of the science of virology which occurred around the midpoint of the twentieth century.

Advances in Techniques

Cell and Tissue Culture. An excellent example of the application of specific techniques is the use of cell and tissue culture in virology. Although viruses, as mentioned, had previously been cultured in cell and tissue preparations, the technique had found limited use because of bacterial contamination. With the availability of antibiotics such as penicillin and streptomycin, viable cultures became easier to maintain. As a result, tissue and cell cultures were widely used in virology from the 1950s onwards.

In 1949 John H. Enders, Thomas H. Weller and Frederick C. Robbins of Harvard showed that poliovirus could be grown in cultures of nonneural tissue. This achievement made possible the development of a vaccine against poliomyelitis.[17] Previous research had been hampered by the fear that a vaccine prepared in brain or nervous tissue cultures might cause a severe reaction in humans.

Enders, Weller and Robbins also found that the presence of a living virus could be detected, without the inoculation of experimental animals, by using a dye which changes color with viral growth in cell and tissue culture. In 1954 they received the Nobel Prize in Physiology and Medicine for their research on poliovirus. By the following year an antipolio vaccine, developed by Jonas E. Salk using the Harvard team's methods, was available for public use.

One consequence of the extensive use of cell and tissue culture techniques was the discovery in the late 1940s and 1950s of many new viruses, such as the myxoviruses, enteroviruses, Coxsackie viruses and ECHO (enteric, cytopathogenic human orphan) viruses. (See Table IV) In 1948 only twenty specifically human viruses were known; by 1959 the number had risen to seventy. (Huebner)

The Plaque Technique. A technique based upon the visible changes produced in host cells by many viruses was introduced in 1952 by Renato Dulbecco, a virologist at the California Institute of Virology. He had used a comparable technique in research on bacteriophage. Areas of focal destruction or 'plaques' produced by certain viruses in cell culture monolayers could be counted with the naked eye if the surrounding 'lawn' of living cells was stained with a vital dye. This plaque assay became an invaluable tool in the titration of viruses and was particularly important in animal virology which heretofore had lacked adequate quantitative methods.[18] In addition, it facilitated research on viral infection, reproduction and immunity at the cellular level.

Electron Microscopy. Another area in which technical improvements facilitated refinement of the concept of the virus was that of electron microscopy. The techniques of ultrathin sectioning and negative staining, the latter introduced by S. Brenner and R.W. Horne in 1959, and mechanical advances in the instrument itself, increased resolving power and led to more precise information about the structure and function of viruses.

All these techniques—cell and tissue culture, the plaque assay, electron microscopy, as well as the previously mentioned x-ray crystallographic, immunological and biochemical methods—decreased the former reliance on pathological changes in the intact organism as the indication of the presence of viruses.

It was learned, furthermore, that a number of viruses isolated in cell and tissue culture had little or no ability to cause disease in the organism from which they were obtained. These are exceptions to the longstanding association of viruses and pathogenicity.

VIRUS GROUPS ISOLATED AND CHARACTERIZED ESSENTIALLY BY TISSUE CULTURE METHODS

Virus	Number of reported serological types	Types of disease which may be produced	References
Enterovirus			
Polioviruses	3	Poliomyelitis, meningo-encephalitis	Weller *et al.* (1949)
ECHO viruses	20	Meningoencephalitis, exanthem	(See Committee on ECHO Viruses, 1955)
Adenovirus	18	Respiratory infections, pharyngitis, conjunctivitis	Hilleman and Werner (1954) Rowe *et al.* (1953)
Measles	1	Specific	Enders and Peebles (1954)
Varicella, zoster	1	Specific	Weller and Stoddard (1952)
Salivary gland virus	1 ?	Generalized disease	Smith (1954)
Other respiratory viruses	3 ?	Minor respiratory infections	Chanock (1956) Chanock *et al.* (1958) Price (1956) Mogabgab and Pelon (1956)
Molluscum contagiosum	1	Specific	Dourmashkin and Febvre (1958)

Table IV The virus discoveries resulting from widespread adoption of tissue culture techniques from the late 1940's onward. (From F.M. Burnet's *Principles of Animal Virology*, 2nd ed., London: Academic Press, 1960.)

A Common Physical Structure

A fundamental conceptual change occurred with the recognition that all viruses share a common structure despite observed differences in morphology, host specificity and pathological activity. Electron microscopic and biochemical studies of the 1950s indicated that a virus possesses an outer coat or 'capsid' of protein and an inner core of nucleic acid, either DNA or RNA, but not both. On the basis of this information viruses for the first time could be accurately defined and classified.

Previous classifications, with the exception of the one suggested by Bawden in 1941, had been based on disease symptoms rather than on the physical characteristics of a particular virus. They had not been entirely satisfactory. (Fenner) In 1962 A. Lwoff, R. Horne, and P. Tournier proposed a "system of viruses" in which viruses were grouped according to their DNA or RNA content, their helical or cubic symmetry, and the naked or enveloped state of the capsid. Their system was the basis for the classification proposed by the Provisional Committee for the Nomenclature of Viruses in 1965. More recently (1974) Fraenkel-Conrat compiled an alphabetical catalogue describing almost all viruses and virus groups of vertebrates, plants, insects and protists.

Virus-Cell Interactions

Replication at the Cellular Level. A second fundamental conceptual change was the realization that all viruses exist in either of two basic forms: the metabolically inert, transmissible virion and the vegetative virus, an active constituent of the host cell. As shown primarily through experiments on bacteriophage, the growth cycle of a virus can be divided into five stages—adsorption of the virion to the cell surface, penetration of the nucleic acid component into the cell, replication of the viral nucleic acid by the infected cell, maturation, and release of virions at the cell surface.[19]

Replication at the Molecular Level. A third fundamental conceptual change which occurred around the middle of the century was that all viruses, indeed all forms of life, share a common mechanism of replication at the molecular level. This mechanism was gradually worked out through contributions from many areas of research.

In 1944 the bacteriologist Oswald Theodore Avery and his associates Colin MacLeod and Maclyn McCarty at the Hospital of the Rockefeller Institute for Medical Research demonstrated that hereditary traits could be transferred from one type of bacterial cell to another by purified fragments of DNA. Since DNA was known to be present in the chromosomes of all cells, these transformation experiments suggested that DNA rather than protein, as currently believed, was the carrier of genetic information.

In 1952 A.D. Hershey and Martha Chase at the Cold Spring Harbor Laboratory of Quantitative Biology provided further support for this belief by demonstrating with radioactively-labeled bacteriophage that only their nucleic acid moiety entered the cell, the protein coat remaining at the cell surface. Hence DNA had to be directing viral replication. Four years later A. Gierer and G. Schramm at the Max-Planck-Institut für Virusforschung in Tübingen and, independently, H. Fraenkel-Conrat of the Virus Laboratory at the University of California in Berkeley, demonstrated that RNA was the infective component of tobacco mosaic virus. Similar discoveries in regard to the DNA of animal viruses were made within the year. The genetic role of viral nucleic acids was established.

By 1953 James D. Watson and Francis H.C. Crick had proposed the double helix model for DNA structure and function based partially on data obtained from x-ray diffraction and biochemical analyses. Their prediction of semiconservative DNA duplication was experimentally confirmed in 1958 by Matthew Meselson and Franklin Stahl in research on bacteriophage. Thus virology, concerned as it is with the subcellular domain, has become intimately linked to the fields of molecular biology and genetics.

After the DNA molecule had been established as the fundamental unit of biological replication, investigators turned with new insight to such problems as genetic replication, RNA function, protein synthesis and genetic mutation.

In 1959 Robert L. Sinsheimer of the California Institute of Technology reported the discovery of a single-stranded form of DNA in a minute bacteriophage. This finding raised the question of whether this single molecule of DNA conformed to the double helix model of Watson and Crick for its replication. Sinsheimer proposed that the single-stranded DNA, upon injection into the cell, assumes an altered "replicative form" with chemical and physical properties "similar to those of double-stranded DNA." It is then multiplied many times following the Watson-Crick model of complementary base pairing. Subsequently it was found that the DNA of Sinsheimer's phage and its replicative form, as well as other viral DNAs, exist in ring form. (Kleinschmidt, Burton and Sinsheimer) A double-stranded RNA[20] was later identified in certain plant and animal viruses.

In 1970 David Baltimore, Alice S. Huang and Martha Stampfer at the Massachusetts Institute of Technology discovered the enzyme RNA polymerase in virions of vesicular stomatitis virus. They suggested that the viral RNA serves as a template for complementary RNA synthesis in the infected cell using the virion's RNA polymerase. This complementary RNA then acts as a messenger for the intracellular translation of viral proteins.

In the same year Baltimore and, independently, Howard M. Temin and Satoshi Mizutaki at the University of Wisconsin found another viral enzyme, this time in virions of RNA tumor viruses. They proposed that the enzyme, an RNA-dependent DNA polymerase, assists in the intracellular synthesis of a DNA copy which becomes a template for the production of viral RNA. Their findings indicate that the classical process of genetic transcription can be reversed: Information can be passed from RNA to DNA, as well as from DNA to RNA. All these investigations, beginning with Sinsheimer's, indicate that there are several molecular models for viral replication.

Viruses continue to hold a primary position in investigations concerning these fundamental molecular processes primarily because of their relative simplicity, both genetically and biochemically, as compared with the cell. As a result virology from the 1950s onwards has been more than ever an interdisciplinary meeting ground.

Lysogeny and Interference. The bonds between virology and molecular biology and genetics were strengthened by the rediscovery in the 1950s of two phenomena associated with viruses: lysogeny and interference.

The term 'lysogenic' was used as early as 1921 by J. Bordet and M. Ciuca of the Pasteur Institute in Brussels to describe bacteria which, in the absence of infection, have the hereditary power to release bacteriophage which lyse sensitive bacteria. However, the mechanism of this occurrence was not understood.

In 1929 F.M. Burnet and M. McKie in Melbourne attributed lysogeny in bacteria to the presence in each cell of an "anlage," an inheritable lytic principle which, they suggested, is a normal component of the genetic constitution of lysogenic bacteria and is potentially capable of being liberated as phage. Although they did now know how this liberation occurred, their evidence suggested that unless the anlage was activated, phage were not released, even upon disruption of the bacterium.

The understanding of lysogeny was extended by the work of A. Lwoff and A. Gutmann of the Pasteur Institute in Paris. They attributed lysogeny to the presence of "probacteriophage" in the bacterial genome which either could be transmitted innocuously to daughter cells or "activated" in some fashion to cause lysis of sensitive bacteria.

The phenomenon of interference also involves an intimate association of virus and cell. The interesting fact that one virus can inhibit the growth of another had been recognized in animal viruses since 1935 (even earlier in plant viruses). Meredith Hoskins reported in that year that when two strains of yellow fever virus, one fatal and the other nonfatal, were administered simultaneously to rhesus monkeys, the nonfatal strain had a protective effect; fewer monkeys died. Hoskins, however, was unable to find "protective substances" in emulsions of infected materials and the observation remained unexplained for many years.

In 1957 A. Isaacs and J. Lindenmann of the National Institute for Medical Research in London attributed one type of interference between viruses to a substance, "interferon," which they found was a protein. It is now known that interferons, induced from infected cells by many types of viruses, inhibit viral synthesis within the host cell. But it remains unclear exactly how induction occurs and also how interferon inhibits the replication of viruses. Although interferon has not seen much use as an antiviral agent, research on the phenomenon and on lysogeny has provided insight into the mechanism of viral replication and the molecular aspects of virus-cell interaction.

Viroids

A deviation from the standard concept of the virus was discovered in the early 1960s when Theodor O. Diener of the

Agricultural Research Service in Beltsville, Maryland, succeeded in extracting an infectious substance from plants with potato spindle tuber disease. At first he suspected a virus but later decided on the basis of density gradient centrifugation studies that the material sedimented far too slowly to be a nucleoprotein. Finding that it was destroyed by ribonuclease and that it had a molecular weight far lower than that of the smallest viruses, he suggested that he was dealing with a free, single-stranded RNA with an unusually low molecular weight. Similar infectious RNAs were found to be the cause of other plant diseases. In 1971 Diener proposed the term 'viroid' meaning 'virus-like' for these free RNAs which are able to reproduce in several genetically different hosts. Viroids, it is believed, must depend for their replication on the biosynthetic systems of the plant since there is no evidence of a 'helper' virus mediating viroid replication.

With the discovery of the viroid, a new dimension was added to the concept of the submicroscopic infectious agent. It now appears that replicating units even smaller and simpler than the virus are able to incite disease in plants and possibly also in animals.

Emergence of Virology as an Independent Science

The emergence of virology as an independent science occurred sometime around the midpoint of the twentieth century. There are several reasons for making this statement. First, as we have seen, viruses had come to be recognized by then as unique infectious agents, distinct from micro-organisms and all other forms of life or nonlife, particularly in their manner of growth. Second, specialized techniques for research on the viruses had been developed. Third, around this time those engaged in research on viruses began to recognize themselves as 'virologists' rather than as pathologists, biochemists, microbiologists, etc. Naturally if there were virologists there had to be a virology—an area of research which was no longer a mere appendage of microbiology, biochemistry or plant and animal pathology.

Fourth, investigators of animal, plant, insect and bacterial viral diseases, recognizing the structural and functional similarities of viruses, focused increasingly on a common interest, the nature of viruses, and less on the diseases they caused. This reorientation of research is reflected in both the title and contents of Salvador Luria's textbook, *General Virology* (1953), in which he discussed some of the features and functions common to all viruses. Previous books had concentrated mainly on the pathological activities of viruses which are confusingly disparate and tend to obscure their basic similarities. Of course the study of specific diseases has continued to remain an important facet of virological research.

In 1955 an additional link between animal and plant virology was established when F.L. Schaffer and C.E. Schwerdt of the Virus Laboratory at the University of California, Berkeley, succeeded in crystallizing an animal virus, poliovirus, for the first time. This achievement occurred twenty years after Stanley had obtained the tobacco mosaic virus in crystalline form, and was further evidence that there was no fundamental difference between the physical structure of animal and plant viruses. It was now possible to prepare both in a relatively pure form for further investigation.

Fifth, although the *Archiv für die Gesamte Virusforschung* had been circulating since 1939, the first journal in English devoted exclusively to virological research, *Virology*, began publication in 1955. It was followed in 1967 by the *Journal of General Virology* and the *Journal of Virology*.

Sixth, institutions devoted exclusively to virological research were established, such as the Virus Laboratory under the direction of Stanley at the University of California at Berkeley (1952), and the Max-Planck-Institut für Virusforschung in Tübingen (1954).

The Present Status of the Microbial and Nonmicrobial Concepts of the Virus

The extended debate over the microbial and nonmicrobial concepts of the virus and over the associated issue of whether it is living or dead lost its force from the 1950s onwards. Today the virologist tends to avoid such questions in favor of more pressing practical problems concerning viruses. The British virologist C.H. Andrewes was aware of this change in emphasis when he remarked in 1953:

> Less is heard now than a few years ago as to what viruses are and where they came from: such speculation is felt to be rather idle

while there are so many new and exciting things to discover about how they produce their effects.

The virus today is conceived of as a metabolically inert, infectious, but not invariably pathogenic, entity which is smaller than a cell but larger than most macromolecules. It possesses an exclusively intracellular mode of reproduction and consists of a protein or lipoprotein coat and one type of nucleic acid. Thus it can be seen that neither the microbial nor the nonmicrobial concept adequately describes the viruses. They cannot logically be placed in either a strictly biochemical or in a strictly biological category; they are too complex to be macromolecules in the ordinary sense and too divergent in their physiology and manner of replication to be conventional living organisms. Discussing these alternatives in an article entitled "The concept of virus," Lwoff (1957) succinctly acknowledged their uniqueness: "Viruses should be considered as viruses becauses viruses are viruses."

It is clear that the microbial and nonmicrobial concepts of the virus, though today no longer viable, have had a formative role in viral research throughout its history. The existence of two ways of looking at viruses, each entailing a different conceptual and, consequently, a different experimental approach, has contributed to the development of the science of virology and to the modern concept of the virus.

Notes

1. Two other papers describing the filterability of the fowl plague agent appeared at this time: E. Centanni and E. Savonuzzi, "La peste avaria I & II, communicazione fatta all'accademia delle scienze mediche e naturali de Ferrara," 1901, March 9 and April 4; and A. Maggiora and G.L. Valenti, "Su una epizooia di tifo essudativo dei gallinacei," *Accad. med. Modena*, June 20, 1901. In both cases papers in German followed: Centanni (1902) and Maggiora and Valenti (1903). For a detailed discussion of the history of research on the fowl plague virus see: Wilkinson and Waterson (1975).
2. Mrowka's thinking was restricted to animal and human viruses; by "animal" he did not intend specifically to exclude plant viruses.
3. The term 'macromolecule' was not introduced until 1922. Even then it was some time before the concept was widely accepted. It was

thought that seemingly large molecules were actually complexes or "micelles" of smaller molecules. See Olby.

4. The hemoglobin molecule is now taken to be an ellipsoid with the dimensions 64×55×50 Å. (Lehmann and Huntsman)
5. From the Greek for 'mantled animals.'
6. Strongyloplasma is a term coined by Lipschütz from the Greek for 'round' and 'body.'
7. For example, Wolbach classified viruses according to their method of transmission—contact with infectious material, intermediate host, and direct entrance into the body through an abrasion or injury. Philibert divided the viruses into the cytolytic—those that usually cause cellular lysis—and the cytocinetic—those that usually cause cellular proliferation.
8. The rickettsiae form a genus of pathogenic and nonpathogenic micro-organisms which are pleomorphic and visible in the light microscope. Like bacteria, they have a cell wall and are sensitive to antibiotics. They can be grown experimentally only on special nutrient substrates containing living tissue and usually behave as obligate intracellular parasites. The few rickettsiae that are pathogenic to man are spread by lice, fleas and ticks, producing diseases such as epidemic typhus, Rocky Mountain spotted fever and Q fever.
9. In 1908 Marchoux had called for a uniform standard of filtration in order to give the 'filterable virus' a scientific basis.
10. Appendix A contains a history of the word 'virus.'
11. In 1902, Joest had stated: "It is possible that many of the unknown [infectious] agents cannot be cultivated artificially, and not just because they place such stringent demands on the nutritive material. This situation occurs not infrequently, particularly in obligate parasites which develop exclusively inside the animal body." (p. 378)
12. It was later found by electron microscopy that the virus of foot and mouth disease is about 0.024 micron in diameter. For examples of the problems involved in measuring viruses and the variability of sizes reported for the same virus, see Rivers. (1932)
13. In actuality Stanley had obtained fibrous 'paracrystals' of tobacco mosaic virus with a two dimensional rather than a three dimensional regularity. As discussed in Chapter 4, naturally occurring 'crystals' of the mosaic virus were first described in 1902 by Ivanovski who, however, did not recognize their true identity.
14. See Luria and Darnell, pp. 186–190, for a description of the one-step growth experiment.

15. Allen discusses the origin and development of molecular biology in Chapter 6 of his book, *Life Sciences in the Twentieth Century*.
16. In 1902 Centanni reported that he had inoculated embryonated eggs with fowl plague virus. Because the results were inconclusive, he dropped the method which was not revived for many years. (pp. 187–188) In 1911, the year of the discovery of the virus of chicken sarcoma, Peyton Rous and J.B. Murphy reported that filtered extracts of these tumors produced lesions on the chorioallantoic membrane of embryonated eggs. The usefulness of this technique for viral research was not appreciated until after Woodruff and Goodpasture's paper appeared.
17. Saul Benison's oral history memoir of T.M. Rivers contains Rivers' account of the development of the polio vaccine, a development with which he was closely associated.
18. For a firsthand account of the development of the plaque technique, see Dulbecco. (1966)
19. The growth cycle of bacteriophage is described in detail by Luria and Darnell, pp. 190–214.
20. RNA is usually single-stranded.

APPENDIX A

A History of the Word 'Virus'

The word 'virus' has undergone many changes in meaning since the time of its Latin origin at least twenty-one centuries ago. Because of the length of this time period, the following history of its successive meanings must be selective.

In discussing the shifts in meaning of a word, one risks implying that they are sudden and definitive occurrences, whereas in fact they usually happen gradually in response to new needs. In most cases it is impossible to state when a given meaning becomes prevalent and of course the older meaning often persists along with the new. A variety of concurrent interpretations is particularly to be expected with a word such as 'virus' which has been used in both scientific and lay contexts throughout its history. Despite the more precise requirements of scientific language, the term has seldom been employed homogeneously even by scientists of any one period. The following account is an attempt to indicate when certain changes in meaning became apparent in scientific and medical writings; it does not imply that other interpretations necessarily disappeared.

Early Use in Latin. It is usual for the person with some knowledge of Latin to assume that the original meaning of 'virus' is 'poison' or 'venom.' Consequently one discovers with some surprise, upon referring to any comprehensive Latin dictionary, that the first and most general meaning is 'slimy liquid' or 'slime' with no suggestion of poison whatsoever. However, an examination of various Latin authors, both scientific and nonscientific, reveals that the word was more often used with a deleterious connotation. The most common meaning was in fact 'poison' or 'venom' as in the following quotation of about 50 A.D. from the Roman encyclopedist, Cornelius Aulus Celsus, in which the ancient and modern meanings of 'virus' meet by coincidence:

> Especially if the dog is rabid, the virus should be drawn out with a cupping glass.

'Virus' used figuratively in the sense of 'virulent or bitter feelings' occurs with some frequency in nonscientific writings, as in this epigram by Martial:

> Lest you should waste your time, keep your virus for those that fancy themselves.

'Virus' even appears in the sense of 'stench' or 'offensive odor,' as in Pliny the Elder's recipe for a deodorant:

> This treatment is done in baths of two parts honey to one of alum. It allays the virus of the axillae and of the sweat. (vol. 9, p. 398)

Another rather restricted use of 'virus' is with the meaning of 'sharp taste,' a meaning apparent in another reference by Pliny:

> . . . resin is believed both to lessen the extreme harshness of a wine and to allay its virus, or in the case of a flat, mild wine, to add a touch of virus. (vol. 4, p. 268)

Thus the word 'virus' appears in the Latin classics in a wide range of contexts, from poetry to practical treatises. Except for the uncommon meaning of 'slimy liquid,' it always signifies a poison in either the figurative or nonfigurative sense. In fact, from the time of its origin up to the present, 'virus' has been most commonly used to designate something menacing to the health.

The Middle Ages and the Renaissance. Medieval scholars adopted the word 'virus' along with most of the classical Latin vocabulary which was the vehicle of academic expression in the West until the nineteenth century. The word served as a synonym for 'poison,' the meanings of 'slimy liquid' and 'unpleasant odor or taste' seldom if ever being used in the Middle Ages.

Saliceto, the Italian surgeon who taught at Bologna in the thirteenth century, was one of the many medieval writers who employed 'virus' to mean a tangible poison. For example, he wrote of the "virus et venenum" sometimes emitted from the crusts formed in scalp diseases. His pupil, Lanfranc of Milan, gave the word a similar interpretation in one of his works on surgery written towards the end of the thirteenth century. The fifteenth century English translation contains in a chapter on "olde woundes" the earliest known appearence of 'virus' in an English text:

> For every olde wounde, hauynge roteness othere [or] virus that is thenne venemy quyter [discharge], other eny other thynge thenne gode quyter nys [is] not y-clepyde [called] wounde, but is y-clepyde ulcus. . . .

The Sixteenth and Seventeenth Centuries. Various treatises on the cause and prevention of infectious diseases, either written originally in English or translated from European sources, appeared in Britain in the sixteenth and seventeenth centuries. In these translations, 'virus,' which continued to be used in works written in Latin, was usually replaced by an English synonym. For example, the translator of a sixteenth century German pharmacopoeia referred to "the venoume of the Patiente" with bubonic plague (Gabelchover) and the English apothecary, William Boghurst, describing the Great Plague of London, write of the "venome" of the pestilence.

The concept of infection was associated with the word 'virus,' perhaps for the first time, in the Latin writings of Athanasius Kircher, the German Jesuit who is reputed to have been the first to suggest the existence of micro-organisms. (Bulloch, p. 17) In his *Scrutinum Physico-medicum* (1658), he referred repeatedly to the "virus pestiferum" and "virus pestilens" or plague virus. He also wrote of the "tarantulae virus," thereby reverting to the older meaning of 'virus' as a poison not specifically associated with infectious disease.

The Eighteenth and Nineteenth Centuries. The word 'virus' was applied in medical works of the eighteenth and early nineteenth centuries to any substance, whether identified or not, which transmitted an infectious disease. Edward Jenner in his classic account of smallpox vaccination (1798), used the term "cow-pox virus" to refer to the pustular lymph which bore the infection and which, as the vaccine, rendered humans immune to smallpox. Similarly, his "variolous virus" was the fluid which he knew transmitted the disease.

Medical writers of the early nineteenth century used 'virus' in two slightly different senses. The first and more general meaning referred to the causative principles of infectious diseases about whose origin, nature and manner of action little was known. As late as 1869 a contributor to a French medical dictionary made the following comment on a definition of 'virus' proposed earlier in the century:

> . . . if we were obliged to define them [viruses], we would have great difficulty and we could not do better than to refer to one of the first additions of the dictionary of Nysten, where we read that the virus is 'a principle unknown in its essence, and inaccessible to our senses; but inherent in some animal humors and able to transmit the disease which has produced it.' I above all like the vagueness, the indecisiveness of this definition which reveals so well the ignorance in which we remain today relative to the nature, the essence, the composition of viruses. (Gallard, 1869)

The second, more specific meaning followed Jenner's usage designating an exudate of a particular infectious disease. In discussions of the pox disease, 'virus' was often used as a synonym for the specific immunizing lymph or vaccine, as in the respective English and French terms "vaccine virus" and "virus-vaccin." For example, the 1858 edition of Dunglison's medical dictionary defined vaccination as "an operation which consists in inserting the vaccine virus under the cuticle [skin]. . . ."

A new dimension was added to the meaning of the word 'virus' in the late 1870s and 1880s after confirmation of the germ theory of disease. Retaining the general meaning of an agent with infectious properties, it came to embrace specific terms for micro-organisms such as bacillus, vibrio and micrococcus, each having a taxonomic significance lacking in the word 'virus.' In this sense Pasteur could refer without ambiguity in 1881 to one and the same micro-organism as "the anthrax virus" and as "the anthrax bacterium" within the space of two paragraphs. Koch a year later used a similar juxtaposition of terms in a paper on tuberculosis:

> All these facts taken together substantiate the claim . . . that, in these bacilli, we have the true virus of tuberculosis.

The word 'virus,' because of its nonspecific connotations, remained a convenient term for referring to infectious agents which had not yet been identified. Thus Pasteur and other French scientists of the late nineteenth century often referred to the causal agent of rabies, which they were unable to isolate or to identify, as "le virus rabique" even though they believed it to be a micro-organism. In short, the term 'virus' as used by bacteriologists of the 1880s and 1890s meant simply 'an agent of infectious disease.' This is the usage in Pasteur's dictum: "Every virus is a microbe." (1890)

The word 'virus' was applied to submicroscopic filterable infectious agents after their recognition in the late 1890s. It continued after the turn of the century to be used as a synonym for 'infectious agent' and also to specify a virulent substance produced by an infected organism. As late as 1907 'virus' was defined in Gould's medical dictionary as "the poison of an infectious disease, especially one found in the secretion or tissues of an individual or animal suffering from an infectious disease."

Ivanovski used the word once in the longer of his two papers of 1892 in which he reported the filterability of the agent of tobacco mosaic disease; he referred to tobacco mosaic as a "virus" disease, meaning that it was caused by bacteria rather than by microscopic fungi. Loeffler and Frosch, in their publication of 1898 on foot and mouth disease, occasionally used 'virus' to refer to the filterable agent of the disease, which they believed to be an extremely minute micro-organism. Beijerinck, writing in the same year on tobacco mosaic disease, employed the word interchangeably with contagium vivum fluidum. He applied both terms to a submicroscopic, noncellular, infectious entity which reproduced exclusively within living cells.

The Twentieth Century. In the early decades of the twentieth century, the agents of what today are identified as viral diseases were referred to by terms specifying either their filterability or invisibility in the light microscope. By using terms such as 'filterable virus' and 'ultrascopic microbe' and a variety of synonyms, investigators were able superficially to distinguish these agents from most familiar micro-organisms. For example, the first review article on viruses, written by Emile Roux in 1903, was entitled "Sur les microbes dits 'invisible.'" Three years later another Frenchman, Paul Remlinger, wrote a review called "Les microbes filtrants."

As investigators acknowledged that filterability and submicroscopic size were not necessarily correlated, 'filterable virus' became more and more the term of preference. However, by the 1930s it was apparent that this designation had drawbacks; the criterion of filterability applied to a wide range of infectious agents with diverse properties. Obviously, the property of filterability was not in itself sufficient to distinguish the viruses from other sorts of filter-passing pathogens. In 1929 Thomas Rivers admitted that the term 'filterable virus' was "misleading and confusing." Nonetheless he condoned it on the grounds of "wide usage" and "general consent." Shortly after this time the tendency to use 'virus' unmodified gained momentum. By 1941 Rivers could state:

> At present most workers speak of viruses instead of filterable viruses, because of confusion caused by the word *filterable*.

The problem was to discover the intrinsic properties of viruses rather than to characterize them in terms of technique-determined ones. By the 1950s, structural studies of viruses with the electron microscope and information about their nucleic acid content provided a meaningful basis for distinguishing viruses from all other types of infectious agents. From this time on, 'virus' was used with a meaning roughly comparable to that given it today. Modern definitions abound but most characterize the virus as an infectious, but not necessarily pathogenic, entity which is usually submicroscopic, which contains a core of either DNA or RNA covered by a protein or lipoprotein capsid and which reproduces exclusively within living cells.

APPENDIX B

Dimitri Iosifovich Ivanovski

The Russian botanist and plant physiologist, Dimitri Iosifovich Ivanovski,[1] was born in the village of Niz near St. Petersburg on October 28, 1864. His father, a commissar of rural police, died when Ivanovski and his four brothers and sisters were young, leaving the family without financial support. The family then moved to a poor section of St. Petersburg where Ivanovski attended secondary school and did tutoring to supplement his mother's small pension. In 1883 he enrolled in the Department of Natural History at the University of St. Petersburg where he was granted free tuition and awarded a stipend to cover his expenses. He chose plant physiology as his speciality, spending the summers of 1887 and 1888 in southern Russia. There under the auspices of the Free Economic Society[2] he investigated diseases of tobacco, including tobacco mosaic disease, on plantations in Bessarabia, the Ukraine and the Crimea.

In 1888 Ivanovski received the degree of Candidate of Science but stayed on at the University to prepare for a career in academic teaching and to work in the botanical laboratory. He later resigned to accept an assistantship in Andrei Famintsyn's laboratory for plant physiology[3] at the Academy of Sciences in St. Petersburg where he continued his research on tobacco mosaic disease. In 1889 he married Evdokia Ivanovna, the daughter of a political exile. Their only child, a son Nicolai, was born in 1890.

Having acquired a reputation as an expert on tobacco mosaic disease, Ivanovski was asked by the Department of Agriculture to study the disease on tobacco plantations in the Crimea. He agreed to postpone his research for his Master's degree and spent the summers of 1890 and 1891 in the field and the winters in the laboratory at the Academy of Sciences. This work provided the material for the two publications on tobacco mosaic disease of 1892 which contain the first description of the filterability of an infectious agent.

In 1895 Ivanovski was awarded the degree of Master of Science on the basis of his thesis entitled "Studies on Alcoholic Fermentation." Soon after he began to lecture at the University on the physiology of lower organisms, emphasizing soil microbes. In 1896, with the understanding that he would obtain a doctoral degree within five years, he was appointed Privatdocent with the assignment of teaching plant anatomy and physiology. He also gave a course on fermentation at the Institute of Technology. In recognition of his expertise in soil microbiology, he was invited to join the editorial board of *Pochvovedenie* (*Soil Science*), founded in 1899.

Having failed to complete his doctoral dissertation within the stipulated period, Ivanovski was transferred in 1901 to Poland where he became associate professor of plant anatomy and physiology at the University of Warsaw. The following year he completed his doctoral dissertation on tobacco mosaic disease and in 1903 accepted the chair in plant anatomy and physiology. From then on he devoted his time to teaching and to research on plant physiology, virtually abandoning his microbiological work of the St. Petersburg period.

In 1911 his son, a student at the University of Moscow, died of tuberculosis, an event which accelerated Ivanovski's already deteriorating health. Another tragedy, the outbreak of World War I, forced the evacuation of the University of Warsaw to Rostov-on-Don in the Ukraine in 1915. In the confusion, Ivanovski lost his library, laboratory notebooks and equipment. He spent the last years of his life writing a textbook of plant physiology which was published in two parts in 1917 and 1919. Suffering from what was perhaps cirrhosis of the liver, Ivanovski died on June 20, 1920 at the age of fifty-six and was buried in Rostov.

Ivanovski is remembered as a dynamic, provocative lecturer and for his liberal views on the right of women to equal educational opportunities. During the Warsaw period, he employed Poles in his laboratory and in his home at a time when the Soviet government was demanding complete Russianization of the University of Warsaw. His Russian biographers describe him as a shy, private man but one popular with his students. They suggest that Ivanovski's reluctance to promote his work may be one reason why his research on tobacco mosaic disease was not given full recognition at home and abroad for many years. However, from World War II on, the Russians have made considerable effort to have Ivanovski recognized as "the founder of virology."[4] The Institute of Virology, Academy of Medical Sciences of the U.S.S.R. is named in his honor and the Academy also awards the D.I. Ivanovski prize for the year's best work in virology.

Martinus Willem Beijerinck

Martinus Willem Beijerinck,[5] the youngest of four children of a tobacconist, was born in Amsterdam on March 16, 1851. Business difficulties forced the family to move to Haarlem where Beijerinck's father secured a job with the Holland Railway. Beijerinck attended the Hoogere Burgerschool in Haarlem and developed such a love for the local flora that he decided upon botany as a career. He graduated in 1869, but lacked money for the university education in botany which he desired. With financial assistance from an uncle, he enrolled in a three year course in technology at the Polytechnical School in Delft. There he received a foundation in chemistry which was to serve him well in his later botanical and microbiological research.

Beijerinck received a diploma in chemical engineering in July 1872 and in October enrolled in the botany program at the University of Leiden. He supported himself as lecturer in botany, physiology and physics at the Agricultural School in Warffum and later at a secondary school in Utrecht while he studied for his doctoral examination. In 1876 he was appointed to the Agricultural School in Wageningen where he taught and did botanical research.

In 1877 he received the doctoral degree for his dissertation on the morphology of plant galls. On the strength of this work he was elected in 1884 at the comparatively young age of thirty-three to the Royal Academy of Sciences of the Netherlands. In the same year he accepted the position of bacteriologist at the Dutch Yeast and Spirit Factory in Delft, although as yet he had had no experience in bacteriology. While the new laboratory promised him was being completed, Beijerinck visited the mycologist Anton de Bary in Strasbourg and the bacteriologist Emil Hansen in Copenhagen in order to learn the microbiological techniques required in his new position. In 1895 he became professor of microbiology at the Polytechnical School in Delft where he founded the laboratory in which he conducted the research which was to establish him as one of the great microbiologists of all time. Beijerinck, who never married, remained in Delft for the next thirty-six years leading a solitary and outwardly uneventful life devoted to scientific pursuits.

Beijerinck received many honors, including the Order of the Dutch Lion (1903) awarded by the Dutch Government, and the

Leeuwenhoek Medal (1905) of the Royal Academy of Sciences of the Netherlands. He also received various awards from abroad and was a member of several foreign scientific and medical societies. On his seventieth birthday in 1921 Beijerinck was presented with five volumes of his scientific papers (*Verzamelde Geschriften*), which had been published through contributions from friends, former students and the Dutch Government. He reluctantly retired a few months later to the town of Gorssel in eastern Holland where he lived with his two spinster sisters. He died of cancer on January 1, 1931 at the age of seventy-nine and was buried at Westerveld.

A taciturn and somewhat acerbic personality, Beijerinck was nonetheless respected for the breadth of his scientific knowledge, his exceptional powers of observation and the originality of his research. He wrote over one-hundred-and-forty papers in the fields of botany, microbiology, chemistry and genetics. His scientific reputation, however, is based primarily on his extensive research in microbiology, a field which in his lifetime included the study of viruses. One of his major achievements was the isolation in 1888 of *Rhizobium leguminosarum*, a bacterium which fixes free nitrogen and causes the formation of root nodules on legume plants. He also studied other micro-organisms, including algae and yeasts, and developed the bacteriological techniques of auxanography and enrichment culture (1894). Perhaps Beijerinck's most lasting achievement was to establish, along with Sergei Winogradsky, the importance of micro-organisms in the cycles of matter, and to demonstrate how each is specialized to perform a particular kind of chemical transformation.

Yet Beijerinck was far from being a single-minded microbiologist. He wrote authoritatively on a variety of botanical subjects (gall, bud and root formation, phyllotaxis, gummosis, cross-breeding experiments), colloid chemistry and enzymology (he was one of the first to suggest the existence of plant enzymes). After his retirement in 1921, he devoted himself almost exclusively to botanical research and maintained a correspondence with microbiologists around the world.

Appendix B: Notes

1. The biography of Ivanovski is based on English translations of the following works in Russian:

G.M. Vaindrakh, "D.I. Ivanovski. (Biographical sketch)" in *On Two Diseases of Tobacco. Tobacco Mosaic Disease*, Moscow: Medgiz, 1949, pp. 5-68. (Translated by J.M. Irvine.)

Y.I. Milenushkin and V.I. Basalkevich, "The discoverer of the world of viruses. Centenary of the birth of Dimitri Iosifovich Ivanovski (1864-1920)," *Vop. Virus* 1964, 9: 521-526. (Translated by B. Haigh.)

A.A. Smorodintsev and K.G. Vasil'ev, "Centenary of the birth of D.I. Ivanovski, the founder of virology," *Tsitologiya* 1964, 6(4): 401-404. (Translated by B. Haigh.)

The entry on Ivanovski in *Bol'shaia Meditinskia Entsiklopediia*, A.N. Bakulev, (ed.), vol. 10, Moscow, 1959. (Translated by M. Winder.)
And on the following paper in English: H. Lechevalier, "Dimitri Iosifovich Ivanovski (1864-1920)," *Bacteriol. Rev.* 1972, 36: 135-145. This paper includes a bibliography of Ivanovski's publications.

2. The Free Economic Society, one of the oldest scientific and economic societies in the world (1776-1917), sought to promote the development of agriculture, industry and trade. Leading Russian scientists participated in its work. It succeeded in solving many economic and agricultural problems in Russia and particularly concerned itself with smallpox vaccination.

3. Andrei Sergeevitch Famintsyn (1835-1913) was one of the first Russian botanists to specialize in plant physiology. His laboratory at the Academy of Sciences was the progenitor of the present Institute of Plant Physiology of the U.S.S.R.

4. For example, see notes 1b and c above and K.S. Koshtoyants, "On the fiftieth anniversary of D.I. Ivanovski's discovery of filterable viruses," *Mikrobiologiya* 1942, 11: 139-148 (Russian).

5. The biography of Beijerinck is based on the following:

L.E. den Dooren de Jong, "Beijerinck, the man" in *Verzamelde Gescriften van M.W. Beijerinck*, G. van Iterson, E.E. den Dooren de Jong and C.J. Kluyver (eds.), the Hague: Martinus Nijhoff, vol. 6, 1940, pp. 3-47.

W. Bulloch, "Martinus Willem Beijerinck—1851-1931," *Proc. R. Soc.* 1932, B, 109: i-iii.

A. Mayer, "Der höllandische Botaniker, Bakteriologe und Biologe M.W. Beijerinck," *Naturwissenschaften* 1931, 19: 302-305.

S.A. Waksman, "Martinus Willem Beijerinck. 1851-1931," *Scient. Mon., N.Y.* 1931, 33: 285-288.

S.S. Hughes, "Beijerinck, Martinus Willem" in *The Dictionary of Scientific Biography*, C.C. Gillispie (ed.), New York: Charles Scribner's Sons, supplementary volume (to be published). This entry contains an extensive bibliography of primary and secondary sources for Beijerinck.

Bibliography

Abbé, E. "Beiträge zur Theorie des Mikroskops und der mikroskopischen Wahrnehmung." *Arch. mikrosk. Anat. EntwMech.* 1873, 9: 413–468. Republished in *Gesammelte Abhandlungen von Ernst Abbé*, Jena: Gustav Fischer, 1904, vol. 1, pp. 45–100.

Ackerknecht, E.H., *Rudolf Virchow: Doctor, Statesman, Anthropologist*, Madison: University of Wisconsin Press, 1953.

Allen, G., *Life Sciences in the Twentieth Century*, New York: John Wiley and Sons, Inc., 1975.

Anderson, T.F., "Electron microscopy of phages" in *Phage and the Origins of Molecular Biology*, J. Cairns, G.S. Stent, and J.D. Watson (eds.), New York: Cold Spring Harbor Laboratory of Quantitative Biology, 1966, pp. 63–78.

Andrewes, C.H., "Introduction: viruses yesterday, today and tomorrow," *Br. med. Bull.* 1953, 9: 169–171, 170.

Andriewsky, P., "L'ultrafiltration et les microbes invisibles. 1. Communication: La peste des poules," *Centbl. Bakt. ParasitKde* 1914, Abt. I, 75: 90–93, 92.

Arloing, S., *Les Virus*, F. Alcan (ed.), Paris: Germer Baillièrè, 1891, p. 61.

Arthur, J.C., "Proof that bacteria are the direct cause of the disease in trees known as pear blight." *Bot. Gaz.* 1885, 10: 343–345.

Avery, O.T., C.M. McLeod and M. McCarty, "Studies on the chemical nature of the substance inducing transformation of pneumococcal types," *J. exp. Med.* 1944, 79: 137–138.

Baltimore, D., A.S. Huang and M. Stampfer, "Ribonucleic acid synthesis of vesicular stomatitis virus, II. An RNA polymerase in the virion," *Proc. Natn. Acad. Sci. U.S.A.* 1970, 66: 572–576.

Baltimore, D., "RNA-dependent DNA polymerase in virions of RNA tumour viruses," *Nature* 1970, 226: 1209–1211.

Bassi, A., *Del Mal del Segno*, translated by P.J. Yarrow, Phytopathological Classics, number 10, G.C. Ainsworth and P.J. Yarrow (ed.), Baltimore: American Phytopathological Society, 1958, p. 10.

Bastian, H.C., *The Beginnings of Life: Being Some Account of the Nature, Modes of Origin and Transformation of Lower Organisms*, 2 vols., London: MacMillan, 1872.

———, *Evolution and the Origin of Life*, London: MacMillan, 1874.

Bawden, F.C., and N.W. Pirie, "The isolation and some properties of liquid crystalline substances from solanaceous plants infected with three strains of tobacco mosaic virus," *Proc. R. Soc.* 1937, B, 128: 274–320.

Beale, L.S., "On the germinal matter of the blood, with remarks upon the formation of fibrin," *Q. Jl. microsco. Sci.* 1864, 4: 47-63, 56.

——, "Microscopical researches on the cattle plague. Report to Her Majesty's Commissioners" in *Third Report to the Commissioners Appointed to Inquire into the Origin and Nature, etc. of the Cattle Plague*, London: Eyre and Spottiswoode, 1866, pp. 129-154.

——, *Disease Germs; Their Supposed Nature*, London: J. Churchill and Sons, 1870.

——, *Disease Germs; Their Real Nature*, London: J. Churchill and Sons, 1870.

——, *Disease Germs; Their Nature and Origin*, London: J. and A. Churchill, 2nd ed., 1872. The two previous books of 1870 were combined and published as one volume in 1872.

——, *Vitality: An Appeal, an Apology and a Challenge*, London: Churchill, 1898.

Béchamp, A., "Du rôle de la craie dans les fermentations butyrique et lactique, et des organismes actuellement vivants qu'elle contient," *C. r. hebd. Séanc. Acad. Sci., Paris* 1866, 63: 451-455, 451.

——, *Les Microzymas dans leur Rapports avec l'Hétérogénie, l'Histogénie, la Physiologie et al Pathologie*, Paris: J.B. Baillière and Sons, 1883, p. 924.

Bechhold, H., "Durchlassigkeit von Ultrafiltern," *Z. phys. Chem.* 1908, 60: 257-318.

Beijerinck, M.W., *Verzamelde Geschriften van M.W. Beijerinck*, G. van Iterson, L.E. den Dooren de Jong and C.J. Kluyver (eds.), 5 vols., the Hague: Martinus Nijhoff, 1921. A sixth volume, published in 1940, contains papers by Beijerinck which appeared after 1921, indices to the six volumes, and biographical material.

——, "Ueber das Cecidium von Nematus Capreae auf Salix amygdalina" (1888) in *ibid.*, vol. 2, pp. 123-137.

——, "L'auxanographie, ou le méthode de l'hydrodiffusion dans la gélatine appliquée aux recherches microbiologiques" (1889) in *ibid.*, vol. 2, pp. 190-193.

——, "Ueber Spirillum desulfuricans als Ursache von Sulfatreduction" (1894) in *ibid.*, vol. 3, pp. 102-127.

——, "Sur la cécidiogénèse et la génération alternante chez le Cynips calicis" (1897) in *ibid.*, vol. 3, pp. 199-232.

——, "Over een contagium vivum fluidum als oorzaak van de vlekziekte der Tabaksbladen," *Versl. gewone Vergad. wis-en natuurk. Afd. K. Akad. Wet. Amst.* 1898, 7: 229-235. A preliminary paper, dated November 26th, 1898.

——, "Ueber ein Contagium vivum fluidum als Ursache der Fleckenkrankheit der Tabaksblätter," *Verh. K. Akad. Wet.* 1898, 6: 3–21. The full account, dated December, 1898. Translation: "Concerning a contagium vivum fluidum as a cause of the spot-disease of tobacco leaves," translated by J. Johnson, *Phytopath. Class.* 1942, 7: 33–52.

——, "Ueber ein Contagium vivum fluidum als Ursache der Fleckenkrankheit der Tabaksblätter," *Centbl. Bakt. ParasitKde* 1899, Abt. I, 5: 27–33. A translation of the Dutch paper of 1898; dated November 19th, 1898.

——, "Bemerkung zu dem Aufsatz von Herrn Iwanowsky über die Mosaikkrankheit der Tabakspflanze," *Centbl. Bakt. ParasitKde* 1899a, Abt. I, 5: 310–311, 310.

——, "De l'existence d'un principe contagieux vivant fluide agent de la nielle des feuilles de tabac," *Archs néerl. Sci.* 1900, series 2, 3: 164–186. Also in *Verzamelde Geschriften, op. cit.*, vol. 3, pp. 296–312. This is also the full account similar to the paper of 1898 in *Verh. K. Akad. Wet.* However, the paper of 1900 includes as a postscript Beijerinck's response in 1899 to Ivanovski.

Beijerinck, biographies of:

W. Bulloch, "Martinus Willem Beijerinck—1851-1931," *Proc. R. Soc.* 1932, B, 109: i–iii.

L.E. den Dooren de Jong, "Beijerinck, the man" in *Verzamelde Geschriften, op. cit.*, vol. 6, pp. 1–47.

A.J. Kluyver, "Beijerinck, the microbiologist" in *Verzamelde Geschriften, op. cit.*, vol. 6, pp. 97–154.

A. Mayer, "Der höllandische Botaniker, Bakteriologe und Biologe M.W. Beijerinck," *Naturwissenschaften* 1931, 19: 302–305.

G. van Iterson, Jr., "Beijerinck, the botanist" in *Verzamelde Geschriften, op. cit.*, vol. 6, pp. 49–96.

S.A. Waksman, "Martinus Willem Beijerinck, 1851-1931," *Scient. Mon., N.Y.* 1931, 33: 285–288.

Belloni, L., *Le "Contagium Vivum" avant Pasteur*, Paris: Imprimerie Alençonnaise, 1961.

Benison, S., *Tom Rivers: Reflections on a Life in Medicine and Science*, Cambridge: M.I.T. Press, 1967.

Boghurst, W., *Loimographia, an Account of the Great Plague of London in the Year 1665*, J.F. Payne (ed.), London: Shaw and Sons, 1894, pp. 10, 48 *et passim*.

Bordet, J., and M. Ciuca, "Déterminisme de l'autolyse microbienne transmissible," *C. r. Séanc. Soc. Biol.* 1921, 84: 276–278, 278.

Bradbury, S., *The Evolution of the Microscope*, Oxford: Pergamon Press, 1967.

Brenner, S., and R.W. Horne, "A negative staining method for high resolution electron microscopy of viruses," *Biochim. biophys. Acta* 1959, 34: 103–110.

Brock, T.D. (ed.), *Milestones in Microbiology*, London: Prentice-Hall International, 1961.

Buchner, H., "Kurze Uebersicht über die Entwickelung der Bacterienforschung seit Naegeli's Eingreifen in dieselbe," *Münch. med. Wchshr.* 1891, 38: 435–437, 454–456, 436.

Buist, J.B., *Vaccinia and Variola. A Study of Their Life History*, London: J. and A. Churchill, 1887.

Bulloch, W., *The History of Bacteriology*, London: Oxford University Press, 1938; 2nd ed., 1960.

Burnet, F.M., and M. McKie, "Observations on a permanently lysogenic strain of *B. enteritidis gaertner*," *Aust. J. exp. Biol. med. Sci.* 1929, 6: 277–284, 283.

Burrill, T.J., "Report on botany and vegetable physiology," *Trans. Ill. St. hort. Soc.* 1877, n.s., 11: 114–116.

———, "Anthrax of fruit trees; or the so-called fire blight of pear, and twig blight of apple trees," *Proc. Am. Assoc. Advmt. Sci.* 1880, 29: 583–597, 597.

Cagniard-Latour, C., "Mémoire sur la fermentation vineuse," *Annls. Chim. Phys.* 1838, 68: 206–222.

Celsus, *De Medicina*, translated by W.G. Spencer, 3 vols., Loeb Classical Library, London: William Heinemann, 1961, p. 112.

Centanni, E., "Die Vogelpest. Beitrag zu dem durch Kerzen filtrierbaren Virus," *Centbl. Bakt. ParasitKde* 1902, Abt. I, *Orig.*, 31: 145–152, 182–201.

Chamberland, C., "Sur un filtre donnant de l'eau physiologiquement pure," *C. r. hebd. Séanc. Acad. Sci., Paris* 1884, 99: 247–248.

Chauveau, A., "Natur du virus vaccin. Détermination expérimentale des éléments qui constituent le principe actif de la serosité vaccinale virulente," *C. r. hebd. Séanc. Acad. Sci., Paris* 1868, 66: 289–293.

Cheyne, W.W. (ed.), *Recent Essays by Various Authors on Bacteria in Relation to Disease*, London: New Sydenham Society, 1886.

Cohn, F., "Untersuchurgen über Bacterien. IV. Beiträge zur Biologie der Bacillen," *Beitr. Biol. Pfl.* 1876, 2: 249–276.

Cornil, A.-V., and V. Babes, *Les Bactéries et leur Role dans l'Etiologie, l'Anatomie et l'Histologie Pathologiques des Maladies Infectieuses*, F. Alcan (ed.), 2 vols., Paris: Germer Baillière, 1885; 2nd ed., 1886; 3rd ed., 1890, vol. 2, p. 285.

Cornil, A.-V., "Discussion. Sur les ptomaines, les leucomaines et la théorie microbienne," *Bull. Acad. Méd.* 1886, 2nd series, 15: 651-656, 656.

Creighton, C., *A History of Epidemics in Britain*, 2 vols., Cambridge: University Press, 1891-1894; 2nd ed., London: Frank Cass, 1965.

Dale, H.H., "The biological nature of the viruses," *Nature, Lond.* 1931, 128: 599-602.

Darlington, C.D., "Heredity, development and infection," *Nature, Lond.* 1944, 154: 164-169.

de Bary, A., *Vorlesungen über Bacterien*, Leipzig: Wilhelm Engelmann, 1885, p. 136.

d'Herelle, F.H., "Sur un microbe invisible antagoniste des bacilles dysentériques," *C. r. hebd. Séanc. Acad. Sci., Paris* 1917, 165: 373-375, 373-374.

Diener, T.O., "Isolation of an infectious, ribonuclease-sensitive fraction from tobacco leaves recently inoculated with tobacco mosaic virus," *Virology* 1962, 45: 140-146.

———, "Potato spindle tuber 'virus.' IV. A replicating, low molecular weight RNA," *Virology* 1971, 45: 411-428.

Dobell, C., *Antony van Leeuwenhoek and His "Little Animals,"* New York: Dover, 1960, p. 220.

Doerr, R., "Ueber filtrierbares Virus," *Centbl. Bakt. ParasitKde* 1911, Abt. I, Ref., 50: 12-23.

———, "Die Entwicklung der Virusforschung und ihre Problemik" in *Handbuch der Virusforschung*, R. Doerr and C. Hallauer (eds.), Vienna: Julius Springer, vol. 1, 1st half, 1938, pp. 1-21.

Doetsch, R.N. (ed.), *Microbiology: Historical Contributions from 1776 to 1908*, New Brunswick: Rutgers University Press, 1960.

Dubos, R.J., *Louis Pasteur: Free Lance of Science*, London: Victor Gallanz. 2nd ed., 1951, pp. 116-158.

Duclaux, E., *Ferments et Maladies*, Paris: G. Masson, 1882.

———, *Le Microbe et la Maladie*, Paris: G. Masson, 2nd ed., 1886.

———, *Traité de Microbiologie*, 4 vols., Paris: G. Masson, 1898-1901.

Dulbecco, R., "Production of plaques in monolayer tissue cultures by single particles of an animal virus," *Proc. natn. Acad. Sci. U.S.A.* 1952, 38: 747-752.

———, "The plaque technique and the development of quantitative animal virology" in *Phage and the Origins of Molecular Biology*, J. Cairns, G.S. Stent and J.D. Watson (eds.), New York: Cold Spring Harbor Laboratory of Quantitative Biology, 1966, pp. 287-291.

Dunglison, R., *A Dictionary of Medical Science*, Philadelphia: Blanchard and Lea, 1858.

Ehrlich, P., "Die Wertbemessung des Diphtherieheilserums und deren theoretische Grundlagen," *Klin. Jb.* 1897, 6: 299–328.

Elford, W.J., "A new series of graded collodion membranes suitable for general bacteriological use, especially in filterable virus studies," *J. Path. Bact.* 1931, 34: 505–521.

Elford, W.J., and C.H. Andrewes, "The sizes of different bacteriophages," *Brit. J. exp. Path.* 1932, 13: 446–456.

Ellermann, V., and O. Bang, "Experimentelle Leukämie bei Hühnern," *Centbl. Bakt. ParasitKde* 1908, Abt. I, 46: 595–609, 608.

Ellis, E.L., and M. Delbrück, "The growth of bacteriophage," *J. gen. Physiol.* 1939, 22: 365–384.

Ellis, E.L., "Bacteriophage: one step growth" in *Phage and the Origins of Molecular Biology*, J. Cairns, G.S. Stent and J.D. Watson (eds.), New York: Cold Spring Harbor Laboratory of Quantitative Biology, 1966, pp. 53–62.

Enders, J.H., T.H. Weller and F.C. Robbins, "Cultivation of the Lansing strain of poliomyelitis virus in cultures of various human embryonic tissues," *Science, N.Y.* 1949, 109: 85–87.

Fenner, F., *The Biology of Animal Viruses*, 2 vols., New York: Academic Press, 1968, p. 6.

Fischer, A., *Vorlesungen über Bakterien*, Jena: Gustav Fischer, 1897. Translation: by A.C. Jones, *The Structure and Functions of Bacteria*, Oxford: Clarendon Press, 1900, p. 138.

———, "Die Bacterienkrankheiten der Pflanzen. Antwort an Herrn Dr. Erwin F. Smith," *Centbl. Bakt. ParasitKde* 1899, Abt. II, 5: 278–287.

Fischer, I. (ed.), *Biographisches Lexicon der Hervorragende Aerzte der Letzten Fünfzig Jahre*, 2 vols., Vienna: Urban and Schwarzenberg, 1932–1933.

Fleming, D., "Emigré physicists and the biological revolution" in *The Intellectual Migration: Europe and America, 1930–1960*, D. Fleming and B. Bailyn (eds.), Cambridge: Harvard University Press, 1969, pp. 152–189.

Flügge, C., *Die Mikroorganismen*, Leipzig: F.C.W. Vogel, 2nd ed., 1886. Translation of the 2nd ed.: *Micro-organisms*, translated by W.W. Cheyne, London: New Sydenham Society, 1890.

Ford, W.W., *Bacteriology*, Clio Medica Series, New York: Hafner, 1964, pp. 45–89.

Foster, W.D., *A History of Parasitology*, London: E. & S. Livingstone, 1965, p. 114.

———, *A History of Medical Bacteriology and Immunology*, London: William Heinemann, 1970, pp. 1–21.

Fraenkel-Conrat, H., "The role of the nucleic acid in the reconstitution of active tobacco mosaic virus," *J. Am. chem. Soc.* 1956, 78: 882–883.

———, *Descriptive Catalogue of Viruses*, vol. 1 of *Comprehensive Virology*, 5 vols., H. Fraenkel-Conrat and R.R. Wagner (eds.), New York: Plenum Press, 1974–1975.

Gabel[c]hover, O., *The Boock of Physicke*, translated into low Dutch by C. Battus, translated into English by "A.M.," Dordrecht: Isaack Caen, 1599, pp. 301-302.

Gallard, T., *Nouveau Dictionnaire de Médecine et de Chirurgie Practiques*, Paris: J.B. Baillière and Sons, 1869, vol. 9, p. 218.

Galloway, I.A., and W.J. Elford, "Filtration of the virus of foot-and-mouth disease through a new series of graded collodion membranes," *Brit. J. exp. Path.* 1931, 12: 407–425.

Gierer, A., and G. Schramm, "Infectivity of ribonucleic acid from tobacco mosaic virus," *Nature, Lond.* 1956, 177: 702–703.

Gillispie, C.C., *Dictionary of Scientific Biography*, New York: Charles Scribner's Sons, 1971, in progress.

Goodheart, C.R., *An Introduction to Virology*, Philadelphia: W.B. Saunders, 1969, pp. 239–249.

Goodpasture, E.W., "Cellular inclusions and the etiology of virus diseases," *Archs Path.* 1929, 7: 114–132.

Gould, G.M., *The Practitioner's Medical Dictionary*, London: H.K. Lewis, 1907.

Green, R.G., "On the nature of filterable viruses," *Science, N.Y.*, 1935, 82: 443–445.

Guarnieri, "Richerche sulla patogenesi ed etiologia dell' infezione vaccinica e variolosa," *Archo Sci. med.* 1893, 16, 403–424. Review by G. Sanarelli of Guarnieri's paper of 1894: "Ueber die Parasiten der Variola und der Vaccine," *Centbl. Bakt. ParasitKde* 1894, Abt. I, 16: 299–300.

Gye, W.E., "The aetiology of malignant new growths," *Lancet* 1925, 2, 109–117.

Hahon, N. (ed.), *Selected Papers on Virology*, Englewood Cliffs, New Jersey: Prentice-Hall, 1964.

Hall, T.S., *Ideas of Life and Matter*, Chicago: University of Chicago Press, 1969, vol. 2, pp. 316–354.

Hartig, R., *Lehrbuch der Baumkrankheiten*, Berlin: Julius Springer, 1882, p. 27.

Heintzel, K.G.E., *Contagiöse Pflanzenkrankheiten ohne Microben unter besonderer Berücksichtigung der Mosaikkrankheit der Tabaksblätter*, doctoral dissertation, Friedrich Alexanders University, Erlangen, Germany, 1900.

Henderson, W., "Notice of the molluscum contagiosum," *Edinb. med. surg. J.* 1841, 56: 213–218.

Henle, J., "Von den Miasmen und Contagien und von den miasmatisch-contagiosen Krankheiten" in *Pathologische Untersuchungen*, Berlin: August Hirschwald, pp. 1–82. Translation: "On miasmata and contagia," translation and introduction by G. Rosen, *Bull. Hist. Med.* 1938, 6: 907–983, 923.

Hershey, A.D., and M. Chase, "Independent functions of viral protein and nucleic acid in growth of bacteriophage," *J. gen. Physiol.* 1952, 36(1): 39–56.

Hershey, A.D., "The injection of DNA into cells by phage" in *Phage and the Origins of Molecular Biology*, J. Cairns, G.S. Stent and J.D. Watson (eds.), New York: Cold Spring Harbor Laboratory of Quantitative Biology, 1966, pp. 100–108.

Hirsch, A. (ed.), *Biographisches Lexicon der Hervorrangenden Aerzte Aller Zeiten und Völker*, 6 vols., Vienna: Urban and Schwarzenberg, 1884–1888.

Hirst, G.K., "The agglutination of red cells by allantoic fluid of chick embryos infected with influenza virus," *Science, N.Y.* 1941, 94: 22–23.

Hoskins, M., "A protective action of neurotropic against viscerotropic yellow fever virus in Macacus rhesus," *Am. J. trop. Med.* 1935, 15: 675–680, 680.

Huebner, R.J., "Implications of 70 newly recognized viruses in man" in *Perspectives in Virology*, M. Pollard (ed.), London: Chapman and Hall, 1959, vol. 1, pp. 121–144, p. 121.

Hueppe, F., *Naturwissenschaftliche Einführung in die Bakteriologie*, Wiesbaden: C.W. Kreidel, 1896, p. 256.

Hunger, F.W.T., "Untersuchungen und Betrachtungen über die Mosaikkrankheit der Tabakspflanze," *Z. PflKrankh.* 1905, 15: 257–311.

——, "Neue Theorie zur Ätiologie der Mosaikkrankheit des Tabaks," *Ber. dt. bot. Ges.* 1905a, 23: 415–418.

Isaacs, A., and J. Lindenmann, "Virus interference. I. The interferon," *Proc. R. Soc.* 1957, B, 147: 258–267.

Iwanowsky (Ivanovski), D.I., and W. Poloftzoff, "Die Pockenkrankheit der Tabakspflanze," *Zap. imp. Akad. Nauk* (*Mém. Acad. imp. Sci. St. Petersburg*) 1890, 7th Series, 37: 1–24.

——, "Über die Mosaikkrankheit der Tabakspflanze," *Isv. imp. Akad. Nauk* (*Bull. Acad. imp. Sci. St. Petersburg*) 1892, n.s., 3: 67–70. Translation: "Concerning the mosaic disease of the tobacco plant," translation and introduction by J. Johnson, *Phytopath. Class.* 1942, 7: 27–30.

Ivanovski, D.I., "On two diseases of tobacco," *Sel'. Khoz. Lêsov.* 1892a, 169(2): 108–121 (Russian). Also in Ivanovski's *On Two Diseases of Tobacco. Tobacco Mosaic Disease*, Moscow: Medgiz, 1949 (Russian). Translation: In S.S. Hughes' *The Origins and Development of the Concept of the Virus in the Late Nineteenth Century*, doctoral dissertation in the History and Philosophy of Science, University of London, 1972, Appendix, extracts from the paper in the foregoing book translated by J.M. Irvine.

Iwanowski (Ivanovski), D.I., "Ueber die Mosaikkrankheit der Tabakspflanze," *Centbl. Bakt. ParasitKde* 1899, Abt. II, 5: 250–254.

———, "Ueber die Mosaikkrankheit der Tabakspflanze," *Centbl. Bakt. ParasitKde* 1901, 7: 148.

———, "Die Mosaik-und die Pockenkrankheit der Tabakspflanze," *Z. PflKrankh.* 1902, 12: 202–203.

———, "Über die Mosaikkrankenheit der Tabakspflanze," *Z. PflKrankh.* 1903, 13: 1–41. Also in Ivanovski's *On Two Diseases of Tobacco. Tobacco Mosaic Disease, op. cit.*

Ivanovski, D.I., *Fiziologiia Rastenni (Plant Physiology)*, Rostov-on-Don, 1917-1919; 2nd ed., Moscow: State Publishing House, 1924 (Russian).

Ivanovski, secondary sources:

In *Bol'shaia Meditinskia Entsiklopediia*, A.N. Bakulev (ed.) Moscow, 1959 (Russian).

K.S. Koshtoyants, "On the fiftieth anniversary of D.I. Ivanovski's discovery of filterable viruses," *Mikrobiologiya* 1942, 11: 139–148 (Russian).

H. Lechevalier, "Dimitri Iosifovich Ivanovski (1864–1920), *Bact. Rev.* 1972, 36: 135–145.

Y.I. Milenushkin and V.I. Basalkevich, "The discoverer of the world of viruses. Centenary of the birth of Dimitri Iosifovich Ivanovski (1864–1920)," *Vop. Virus* 1964, 9: 521–526 (Russian).

A.A. Smorodintsev and K.G. Vasil'ev, "Centenary of the birth of D.I. Ivanovski, the founder of virology," *Tsitologiya* 1964, 6(4): 401–404 (Russian).

G.M. Vaindrakh, in Ivanovski's *On Two Diseases of Tobacco. Tobacco Mosaic Disease, op. cit.*, pp. 5–68 (biography) and pp. 78–98 (bibliography) (Russian).

Jenner, E., *An Inquiry into the Causes and Effects of the Variolae Vaccinae 1798*, facsimile reprint, London: Dawsons, 1966, pp. 49, 50.

Joest, E., "Unbekannte Infektionsstoffe," *Centbl. Bakt. ParasitKde* 1902, Abt. I, 31: 361–384, 410–422.

Kausche, G.A., E. Pfankuch and H. Ruska, "Die Sichtbarmachung von pflanzlichem Virus in Übermikroskop," *Naturwissenschaften* 1939, 27: 292–299.

Kircher, A., *Scrutinum Physico-medicum Contagiosae Luis, quae Pestis Dicitur*, Rome: Mascardus, 1658, pp. 61, 112, 161, 213, *et passim*.

Kitasato, S., "Experimentelle Üntersuchungen über das Tetanusgift," *Z. Hyg. InfektKrankh.* 1891, 10: 267–305. A description and trial of the Kitasato filter is included.

Klebs, A.C., "The historic evolution of variolation," *Bull. Johns Hopkins Hosp.* 1913, 24: 69–83.

Klein, E., "Infectious diseases, their nature, cause, and mode of spread," *Nature, Lond.* 1891, 43: 416–419, 443–446, 418.

Kleinschmidt, A.K., A. Burton and R.L. Sinsheimer, "Electron microscopy of the replicative form of the DNA of the bacteriophage ØX174," *Science, N.Y.* 1963, 142: 961.

Koch, R., "Untersuchungen über Bacterien. V. Die Aetiologie der Milzbrand-Krankheit, begründet auf die Entwicklungsgeschichte des Bacillus Anthraxis," *Beitr. Biol. Pfl.* 1877, 2: 277–310. Translation: "Investigations of bacteria. The etiology of anthrax, based on the ontogeny of the anthrax bacillus" in *Med. Class.* 1938, 2: 787–820.

———, "Zur Untersuchung von pathogenen Organismen," *Mitt. K. GesundhAmte* 1881, 1: 1–48. Translation: "On the investigation of pathogenic organisms," translated by V. Horsley, in *Recent Essays by Various Authors on Bacteria in Relation to Disease*, W.W. Cheyne (ed.), London: New Sydenham Society, 1886, pp. 1–64, p. 14.

———, "Die Aetiologie der Tuberkulose," *Berl. klin. Wschr.* 1882, 19: 221–230. Translation: "The etiology of tuberculosis," translated by W. de Rouville, in *Med. Class.* 1938, 2: 853–880, 875.

———, "Die Aetiologie der Tuberkulose," *Mitt. K. GesundhAmte* 1884, 2: 1–88. Translation: "The etiology of tuberculosis" (1884), translated by S. Boyd, in *Recent Essays by Various Authors on Bacteria in Relation to Disease*, W.W. Cheyne (ed.), *op. cit.*, pp. 65–201.

———, "Conferenz zur Erörterung der Cholerafrage," *Berl. klin. Wschr.* 1884, 21: 478–483. Translation: "The etiology of cholera," translated by G.L. Laycock, in *Recent Essays by Various Authors on Bacteria in Relation to Disease*, W.W. Cheyne (ed.), *op. cit.*, pp. 325–369, p. 352.

———, "Ueber bacteriologische Forschung," *Deut. med. Wschr.* 1890, 16: 756–757, 756.

Kohler, R., "The background to Eduard Buchner's discovery of cell-free fermentation," *J. Hist. Biol.* 1971, 4: 35–61, 42.

Koning, C.J., *Hollandsche Tabak*, Bussum, 1898 (reprinted from *de Natuur*, 1897).

——, "Die Flecken-oder Mosaikkrankheit des holländischen Tabaks," *Z. PflKrankh.* 1899, 9: 65-80.

——, *Der Tabak. Studien über seine Kultur und Biologie*, Amsterdam: van Heteren, 1900, p. 79.

Kützing, F., "Microscopische Untersuchungen über die Hefe und Essigmutter nebst mehreren andern dazu gehörigen vegetablischen Gebilden," *J. prakt. Chem.* 1837, 11: 385-409.

Lanfrank's "Science of Cirurgie," Early English Text Society, Original Series 102, New York: C. Scribner, 1894, p. 77 (Reprint of the 15th century English text).

Lechevalier, H.A., and M. Solotorovsky, *Three Centuries of Microbiology*, New York: McGraw-Hill, 1965; 2nd ed., New York: Dover, 1974, p. 40.

Lehmann, H., and R.G. Huntsman, *Man's Haemoglobins*, Philadelphia: J.B. Lippincott, 2nd ed., 1974, p. 59.

Leicester, H.M., *The Historical Background of Chemistry*, New York: John Wiley and Sons, 1965, p. 207 and p. 209.

Levaditi, C., "Virus de la poliomyélite et culture des cellules *in vitro*," *C. ren. Séanc. Soc. Biol.* 1913, 75: 202-205.

Levy, E., and F. Klemperer, *Grundriss der Klinischen Bakteriologie. Für Aerzte und Studirende*, Berlin: August Hirschwald, 1894; 2nd ed., 1898, p. 335.

Lipschütz, B., "Ueber mikroskopisch sichtbare, filtrierbare Virusarten (Ueber Strongyloplasmen)," *Centbl. Bakt. ParasitKde* 1909, Abt. I, 48: 77-90.

——, "Filtrierbare Infektionserreger" in *Handbuch der Pathogen Mikroorganismen*, W. Kolle and A. von Wasserman (eds.), Jena: Gustav Fischer, 2nd ed., 1913, vol. 8, pp. 345-426, 351-353.

Lister, J., "On a new method on treating compound fracture, abscess, etc. with observations on the conditions of suppuration," *Lancet* 1867, 1: 326-329, 357-359, 387-389, 507-509; 2: 95-96.

Lode, A., and J. Gruber, "Bakteriologische Studien über die Ätiologie einer epidemischen Erkrankung der Hühner in Tirol (1901)," *Centbl. Bakt. ParasitKde* 1901, Abt. I, 30: 593-604.

Lode, A., "Notizen zur Biologie des Erregers der Kyanolophie die Hühner," *Centbl. Bakt. ParasitKde* 1902, Abt. I, Orig., 31: 447-451, 451.

Loeffler, (-) and (-) Frosch, "Berichte der Kommission zur Erforschung der Maul-und Klauenseuche bei dem Institut für Infektionskrankheiten in Berlin," *Centbl. Bakt. ParasitKde* 1898, Abt. I, 23: 371-391.

Loeffler, (-), "Bericht der Kommission zur Erforschung der Maul-und Klauenseuche bei dem Institut für Infektionskrankheiten in Berlin," *Dt. med. Wschr.* 1898, 24: 562–564.

——, "Bericht über die Untersuchungen der Königlichen Preussischen Commission zur Erforschung der Maul-und Klauenseuche in den Etatsjahren 1901 und 1902," *Dt. med. Wschr.* 1903, 29: 670–672, 685–687.

Loeffler, F., "Ueber filtrierbares Virus," *Centbl. Bakt. ParasitKde* 1911, Abt. I, Ref., 50: 1–12.

Luria, S.E., *General Virology*, London: John Wiley and Sons, 1953; 2nd edition (with J.E. Darnell), 1967.

Lwoff, A., and A. Gutmann, "Recherches sur un *Bacillus megatherium* lysogène," *Annls Inst. Pasteur, Paris* 1950, 78: 711–739, 734.

Maggiora, A., and G.L. Valenti, "Ueber eine Seuche von exsuditivem Typhus bei Hühnern," Z. *Hyg. InfektKrankh.* 1903, 42: 185–243.

Magnin, A., *Les Bactéries*, Paris: F. Savy., 1878.

Marchoux, E., "Bougie filtrantes et virus invisibles," *C. r. Séanc. Soc. Biol.* 1908, 65: 82–84.

Martial, *Epigrams*, translated by W.C.A. Ker, Loeb Classical Library, London: Heinemann, 1961, vol. 2, p. 322.

Mayer, A., "Ueber die Mosaikkrankheit des Tabaks," *Landwn. VersStnen.* 1886, 32: 451–467. First published in Dutch in 1885. Translation: "Concerning the mosaic disease of tobacco," by J. Johnson, *Phytopath. Class.* 1942, 7: 11–24.

McFadyean, J., "African horse-sickness," *J. comp. Path. Ther.* 1900, 13: 1–20.

——, "The ultravisible viruses," *J. comp. Path. Ther.* 1908, 21: 58–68, 168–175, 232–242.

Meselson, M., and F.W. Stahl, "The replication of DNA in *Escherichia coli*," *Proc. natn. Acad. Sci. U.S.A.* 1958, 44: 671–682.

Metchnikoff, E., "Über eine Sprosspilzkrankheit der Daphnien. Beitrag zur Lehre über den Kampf der Phagocyten gegen Krankheitserregers," *Virchows Arch. path. Anat. Physiol.* 1884, 96: 177–195.

Mrowka, (-), "Das Virus der Hühnerpest ein Globulin," *Centbl. Bakt. ParasitKde* 1912, Abt. I, Orig., 67: 249–268, 251.

Negri, A., "Beitrag zum Studium der Aetiologie der Tollwuth," Z. *hyg. InfektKrankh.* 1903, 43: 507–528.

Nocard (-), (-) Roux, (-) Borrel, (-) Salimbeni and (-) Dujardin-Beaumetz, "Le microbe de la péripneumonie," *Annls Inst. Pasteur, Paris* 1898, 12: 240–262.

Nordtmeyer, H., "Ueber Wasserfiltration durch Filter aus gebrannter Infusorienerde," *Z. Hyg. InfektKrankh.* 1891, 10: 145–154.

Nuttall, G., "Experimente über die bacterienfeindlichen Einflüsse des thierischen Körpers," *Z. Hyg. InfektKrankh.* 1888, 4: 353–394.

Olby, R., "The macromolecular concept and the origins of molecular biology," *J. chem. Educ.* 1970, 47: 168–174.

Oppenheimer, C., *Die Fermente und ihre Wirkungen*, Leipzig: F.C.W. Vogel, 1900. Translation: *Ferments and Their Actions*, translated by C.A. Mitchell, London: Charles Griffin, 1901, p. 31.

Partington, J.R., *A Short History of Chemistry*, New York: Harper and Brothers, 3rd ed., 1960.

Pasteur, L., *Oeuvres de Pasteur*, P. Vallery-Radot (ed.), Paris: Masson, 7 vols., 1922–1939.

———, and (-) Joubert, "Étude sur le maladie charbonneuse," *C. r. hebd. Séanc. Acad. Sci., Paris* 1877, 84: 900–906.

Pasteur, L., "Sur les maladies virulentes, et en particulier sur la maladie appelée vulgairement choléra des poules," *C. r. hebd. Séanc. Acad. Sci., Paris* 1880, 90; 239–248, 246.

———, "De l'atténuation des virus et de leur retour à la virulence" (1881) in *Oeuvres de Pasteur, op. cit.*, vol. 6, pp. 332–338, pp. 332–333.

———, "La vaccination charbonneuse, Réponse à un mémoire de M. Koch" (1883) in *Oeuvres de Pasteur, op. cit.*, vol. 6, pp. 418–440, 432.

———, "Microbes pathogènes et vaccins" (1884) in *Oeuvres de Pasteur, op. cit.*, vol. 6, pp. 590–602, p. 600.

———, "Discussion sur les microzymas" (1886) in *Oeuvres de Pasteur, op. cit.*, vol. 7, pp. 67–69, p. 69.

———, "La rage" (1890) in *Oeuvres de Pasteur, op. cit.*, vol. 6, pp. 672–688, p. 673.

Paterson, R., "Cases and observations on the molluscum contagiosum of Bateman, with an account of the minute structure of the tumours," *Edinb. med. Surg. J.* 1841, 56: 279–288.

Philibert, A., "Virus cytotropes (virus filtrants.—virus filtrables)," *Annls Méd.* 1924, 16: 283–308, 307.

Pliny, *Natural History*, translated by W.H.S. Jones, Loeb Classical Library, London: William Heinemann, 10 vols., 1956.

Poggendorff, J.C. (ed.), *Biographisch-literarisches Handwörterbuch zur Geschichte der exacten Wissenschaften*, Amsterdam: B.M. Israël, 6 vols., 2nd ed., 1965.

Remlinger, P., "Les microbes filtrants," *Bull. Inst. Pasteur, Paris* 1906, 4: 337–345, 385–392.

Rivers, T.M., "Some general aspects of filterable viruses," in *Filterable Viruses*, T.M. Rivers (ed.), London: Baillière, Tindall and Cox, 1928, pp. 3-52, pp. 4-5.

——, "Viruses," *J. Am. med. Assoc.* 1929, 92: 1147-1152, 1147.

——, "The nature of viruses," *Physiol. Rev.* 1932, 12: 423-452, 440.

——, "Viruses and Koch's postulates," *J. Bact.* 1937, 33: 1-12, 5.

——, "Virus infections," *Bull. N.Y. Acad. Med.* 1941, 17: 245-258, 247.

Rosen, G., *A History of Public Health*, New York: MD Publications, 1958, pp. 296-318.

Rous, P., "Transmission of a malignant new growth by means of a cell-free filtrate," *J. Am. med. Assoc.* 1911, 56: 198.

Rous, P., and J.B. Murphy, "Tumor implantations in the developing embryo," *J. Am. med. Assoc.* 1911, 56: 741-742.

Roux, E., and A. Yersin, "Contribution a l'étude de la diphtérie," *Annls. Inst. Pasteur, Paris* 1888, 2: 629-661.

Roux, E., "Sur les microbes dits 'invisibles,'" *Bull. Inst. Pasteur, Paris* 1903, 1: 7-12, 49-56.

(Saliceto) de Salicet, G., *Chirurgie*, 1275, translation and commentary by P. Pifteau, Toulouse: Imprimerie Saint-Cyprien, 1898, p. 18.

Sanarelli, G., "Das myxomatogene Virus. Beitrag zum Studium der Krankheitserreger ausserhalb des Sichtbaren," *Centbl. Bakt. ParasitKde* 1898, Abt. I, 23: 865-873.

Schaffer, F.L., and C.E. Schwerdt, "Crystallization of purified MEF-1 poliomyelitis virus particles," *Proc. natn. Acad. Sci. U.S.A.* 1955, 41: 1020-1023.

Schaudinn, F.R., and E. Hoffmann, "Vorläufiger Bericht über das Vorkommen von Spirochaeten in syphilitischen Krankheitsprodukten und bei Papillomen," *Arb. K. GesundhAmt.* 1905, 22: 527-534.

Schleichert, F., *Das Diatatische Ferment der Pflanzen*, Halle, 1893 (cited in Heintzel, p. 43).

Schlesinger, M., "The Feulgen reaction of the bacteriophage substance," *Nature, Lond.* 1936, 138: 508-509.

Schultz, E.W., "The ultramicroscopic viruses from the biological standpoint," *Scient. Mon., N.Y.* 1930, 31: 422-433.

Schwann, T., "Vorlaüfige Mittheilung betreffend Versuche über die Weingährung und Fäulniss," *Annls. Phys.* 1837, 41: 184-193.

Sinsheimer, R.L., "A single-stranded deoxyribonucleic acid from bacteriophage ØX184," *J. molec. Biol.* 1959, 1: 142-160, 142.

——, "ØX: multum in parvo," in *Phage and the Origins of Molecular Biology*, J. Cairns, G.S. Stent and J.D. Watson (eds.), New York:

Cold Spring Harbor Laboratory of Quantitative Biology, 1966, pp. 258–264.

Smith, E.F., "Peach yellows and peach rosette," *Fmrs' Bull. U.S. Dep. Agric.* 1894, no. 17, p. 10.

———, "Are there bacterial diseases of plants? A consideration of some statements in Dr. Alfred Fischer's Vorlesungen über Bakterien," *Centbl. Bakt. ParasitKde* 1899, Abt. II, 5: 271–278, 274.

———, "Dr. Alfred Fischer in the rôle of pathologist," *Centbl. Bakt. ParasitKde* 1899a, Abt. II, 5: 810–817.

———, "Entgegnung auf Alfred Fischer's 'Antwort' in betreff der Existenz von durch Bakterien verursachten Pflanzenkrankheiten," *Centbl. Bakt. ParasitKde* 1901, Abt. II, 7: 88–100, 128–139, 190–199.

———, *Bacteria in Relation to Plant Diseases*, Washington, D.C.: Carnegie Institution, 3 vols., 1905, 1911, 1914.

Stanley, W.M., "Isolation of a crystalline protein possessing the properties of tobacco-mosaic virus," *Science, N.Y.* 1935, 81: 644–645, 645.

———, "Properties of viruses," *Medicine, Baltimore* 1939, 18: 431–442, 442.

Steinhardt, E., C. Israeli and R.A. Lambert, "Studies on the cultivation of the virus of vaccinia," *J. infect. Dis.* 1913, 13: 294–300.

Svedburg, T., and H.R. Rinde, "The ultracentrifuge, a new instrument for the determination of size and distribution of size of particle in amicroscopic colloids," *J. Am. chem. Soc.* 1924, 46: 2677–2693.

Théodoridès, J., "Un grand médecin et biologiste: Casimir-Joseph Davaine (1812–1882)," *Analecta Medico-Historica* 1968, 4: 72–120.

Temin, H.M., and S. Mizutaki, "RNA-dependent DNA polymerase in virions of Rous sarcoma virus," *Nature* 1970, 226: 1211–1213.

Twort, F.W., "An investigation on the nature of ultramicroscopic viruses," *Lancet* 1915, 2: 1241–1243, 1242.

Tyndall, J., "The optical deportment of the atmosphere in relation to the phenomena of putrefaction and infection," *Phil. Trans. R. Soc.* 1876, 166(1): 27–74, 44–45.

Villemin, J.A., *Études sur la Tuberculose; Preuves Rationelles et Expérimentales de sa Spécificité et de son Inoculabilité*, Paris: J.B. Baillière, 1868.

Virchow, R., "Cellular-Pathologie," *Virchows Arch. Path. Anat. Physiol.* 1855, 8: 1–15.

———, *Die Cellularpathologie in ihrer Begründung auf Physiologische und Pathologische Gewebelehre*, Berlin: A. Hirschwald, 1858. Translation: *Cellular Pathology as Based upon Physiological and Pathological Histology*, translated by F. Chance, London: J. Churchill, 1860.

——, "Der Kampf der Zellen und der Bakterien," *Virchows Arch. path. Anat. Physiol.* 1885, 101: 1-13, 10.

——, "Die neueren Fortschritte in der Wissenschaft und ihr Einfluss auf Medicin und Chirurgie," *Berl. klin. Wschr.* 1898, 35: 897-900, 928-934. Translation: "The Huxley lecture on recent advances in science and their bearing on medicine and surgery," *Br. med. J.* 1898, 2: 1021-1028, 1027.

(von) Prowazek, S., "Chlamydozoa. I. Zusammenfassende Übersicht," *Arch. Protistenk.* 1907, 10: 336-358.

Watson, J.D., and F.H.C. Crick, "A structure for deoxyribose nucleic acid," *Nature, Lond.* 1953, 171: 737-738.

Welch, W.H., "General considerations concerning the biology of bacteria, infection and immunity" (1894) in *Papers and Addresses by William Henry Welch*, W.C. Burket (ed.), Baltimore: Johns Hopkins Press, 1920, vol. 2, pp. 3-78.

Wilkinson, L., "The development of the virus concept as reflected in corpora of studies on individual pathogens. 1. Beginnings at the turn of the century," *Med. Hist.* 1974, 17: 211-221.

Wilkinson, L., and A.P. Waterson, "The development of the virus concept as reflected in corpora of studies on individual pathogens. 2. The agent of fowl plague—a model virus?", *Med. Hist.* 1975, 19: 52-72.

Wilkinson, L., "The development of the virus concept as reflected in corpora of studies on individual pathogens. 3. Lessons of the plant viruses—tobacco mosaic virus," *Med. Hist.* 1976, 20: 111-134, 119.

Williams, R.C., and R.W.G. Wyckoff, "The thickness of electron microscopic objects," *J. appl. Physiol.* 1944, 15: 712-716.

Wolbach, S.B., "The filterable viruses, a summary," *J. med. Res.* 1912, 27: 1-25, 22.

Woodruff, C.E., and E.W. Goodpasture, "The infectivity of isolated inclusion bodies of fowl-pox," *Amer. J. Path.* 1929, 5: 1-9.

Woodruff, A.M., and E.W. Goodpasture, "The susceptibility of the chorio-allantoic membrane of chick embryos to infection with the fowl-pox virus," *Am. J. Path.* 1931, 7: 209-222.

Woods, A.F., "The destruction of chlorophyll by oxidizing enzymes," *Centbl. Bakt. ParasitKde* 1899, Abt. II, 5: 745-754.

——, "Observations on the mosaic disease of tobacco," *Science, N.Y.* 1901, n.s., 13: 247-248, 248.

——, "Observations on the mosaic disease of tobacco," *Bull. Bur. Pl. Ind. U.S. Dep. Agric.* 1902, 18: 13-16.

Index

About the Author

Sally Hughes has a Bachelor of Arts degree in zoology from the University of California, Berkeley and a Master of Arts degree in anatomy from the University of California, San Francisco. For several years she did electron microscopical and biochemical research in the Cardiovascular Research Institute at the University of California Medical Center, San Francisco. She then moved to London for three years where she did research on the history of virology, earning her doctorate in the history of medicine at the University of London in 1972. She now lives in Chapel Hill with her physician-husband and three small children.

Library of Congress Cataloging in Publication Data

Hughes, Sally Smith.
The virus.

Bibliography: p.
1. Virology--History. I. Title. [DNLM: 1. Virology--History. 2. Virology--Biography. QW11.1 H893v]
QR359.H84 616.01'94'09 77-512
ISBN 0-88202-168-0